new concepts in renovating

Tadao Ando   Helfand Myerberg Guggenheimer   Rataplan
de Architectengroep (Bjarne Mastenbroek)   Eduardo Souto de
Moura & Humberto Vieria   Vicen Cornu & Benoit Crepet   Gunter
Domenig   Ramón Esteve   Luigi Ferrario   Architekturbüro
Gasparin & Meier   Crone Nation Architects   Jean-Paul Philippon
Klaus Block Architekt   Roberto Luna / Arata Isozaki   Rudy
Ricciotti   José Paulo Dos Santos   Renzo Piano Building
Workshop   Günther Domenig   Klaus Sill & Jochen Keim   Toni
Cordero   Guillermo Vázquez Consuegra   Francesco Delogu &
Gaetano Lixie   Louis Kloster   Sudau, Storch & Ehlers   Adolf
Krischanitz   Adrien Fainsilber & Associates   Erick van Egeraat
Gerhard P. Wirth   Ignacio Mendaro Corsini   Manuel de las Casas
    Prof. Jürg Steiner   Benoîte Doazan & Stéphane Hirschberger,
architectes   Correa + Estévez , arquitectos (Maribel Correa y
Diego Estévez)   Guido Canali   Stéphane Beel & Lieven
Achtergael   Cristian Cirici & Carles Bassó   Aneta Bulant
Kamenova & Klaus Walizer   Jestico + Whiles   Roberto Menghi
Stig L. Andersson   Ottorino Berselli & Cecilia Cassina   Luis Vicente
Flores   Alois Peitz   Claesson Koivisto   Rune Massimiliano
Fuksas Architetto

new concepts *in* renovating

Work concept: Carles Broto
Publisher: Arian Mostaedi

Graphic design & production: Pilar Chueca & Jorge Carmona
Text: Contributed by the architects, edited
        by Jacobo Krauel and Amber Ockrassa

© Carles Broto i Comerma
Jonqueres, 10, 1-5
08003 Barcelona, Spain
Tel.: +34 93 301 21 99 Fax: +34-93-301 00 21
E-mail: info@linksbooks.net

new
concepts
in
renovating

structure

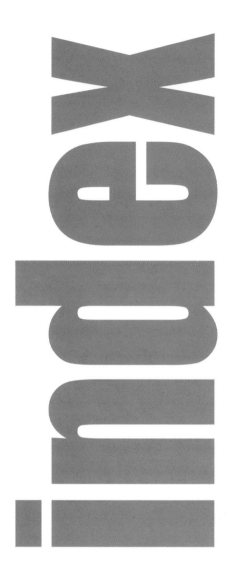

When it comes to renovating, there is no set of "right" or "wrong" criteria. Each project brings with it a unique combination of challenges, problems, strengths and weaknesses which have possibly never been seen before.

On historic buildings, how much of the old should be conserved? How far should the renovation either imitate or diverge from the original? What sort of new technologies and materials are compatible with old structures and finishes? These are just some of the questions which inevitably arise in renovating; and the best architects understand that the answers that apply in one project can never be re-used in subsequent programs. Everything must be reevaluated in light of the new challenges posed by new projects.

The results of our search for some of the most exemplary work currently seen in the field of renovating are varied. Defunct factory buildings, centuries-old stone structures and elegant vaulted spaces are but some of the challenges facing the designers in this collection - all resolved with skill and artistry. From Arata Isozaki and Roberto Luno's masterful reworking of a modernist-era factory to the futuristic Lowe Office buildings by Erick van Egeraat, a wide range of spaces and solutions is presented.

With this selection, we also hope to present a well-rounded vision of each project. To meet this end, we have endeavored to touch upon every aspect in the planning and renovation processes. After all, technical know-how is just as important as artistic vision in any project.

From conception to completion, we have included information on the materials used and construction processes in order to complement the ideas of the contributing architects. Finally, since nobody is in a better position to comment on these projects than the designers themselves, we have included the architects' own comments, conceptual inspiration and anecdotes.

Therefore, we trust that we are leaving you in good, expert hands and that this selection of some of the finest, most innovative architectural solutions in the world will serve as an endless source of inspiration. Enjoy!

Tadao Ando
Benetton Research Center

Treviso, Italy

This 17th century Palladian villa stands in a suburb of Treviso, a city 30 kilometers from Venice in northern Italy. Restoring this villa was the point of departure for the design of a new art school called *Fabrica* sponsored by the Benetton Group. It will invite young students from around the world with achievement in realms such as architectural design, photography, graphic art, image media and textiles to explore and create new forms of and uses for the arts, technology and mass media. Through this research center, students coming together from various international backgrounds exchange their cultures.

The Japanese architect Tadao Ando wanted to express this spirit, engaging his serene concrete architecture and the style of the old Italian villa.

The new additions were sought to bring out the old villa's charm and vitality, and induce —within an overall harmony— a mutually catalytic relationship between the old and the new.

The old villa was restored and converted into studios, a lecture hall, a document center and laboratories. The concrete volume of the elevator shaft and the wide curved concrete auditorium wall with its transparent hinged element stand out against the existing volume. The newly-added columns cross the pond and penetrate the old building, forming a strong statement against the flat Venetian countryside.

The new addition is mostly below ground. A spiral-shaped library, an art gallery and other workshops are located around a subterranean rotunda or along a sunken court.

A new colonnaded gallery is built across the site, penetrating the old villa. On both sides, a large pond welcomes the visitor and creates an effective scenery with the reflection of the villa and its columns on the water.

Photographs: Pino Musi & Tadao Ando

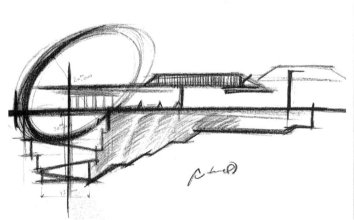

The linear sequence created by the columns reflected on the surface of the water contrasts with the flatness of the Venetian rural landscape and evokes the tranquility of the surroundings.

The addition of new architectural elements gives the old building new features and places in a setting of total harmony.

Helfand Myerberg Guggenheimer
Architects' Office

New York, USA

Representing the first collaboration of a new design partnership, the renovated 6000 ft², 13-foot-high industrial loft space, located in Soho, embodies the design principles of the firm: functional architecture rooted in the traditions of modernism that exploits simple geometry and ordinary materials to poetic effect. The program called for 30 workstations, with extensive technical and material libraries, production areas and meeting spaces of various sizes. The result is an office design which synthesizes a lively, humane working environment while providing many different kinds of experiences, from private endeavor to group collaboration.

Slotted cleanly into the tall industrial space, new elements are simply and crisply expressed. Ordinary materials are used or finished in unexpected ways, animating and enlivening the straightforward layout. The free standing wall set parallel with the line of existing Corinthian columns demarcates entrance and support spaces from more general working areas and provides a remarkably effective sound sink, creating a tranquil and quiet working environment. Workstations and partitions are created from panelled oriented strand board (OSB), finished with aluminum dust. Recycled ground rubber sheeting is used to clad the production screen walls, providing additional sound absorption and tack space. Existing strip flooring is given a luminous finish with bronze dust embedded in polyurethane.

Individual workstations are clustered together in four person pods, strung along a row of north-facing windows. The space between the workstations and dividing wall becomes a processional gallery for project material, leading to a large conference room at the elbow between the two wings. The partners' offices are organized behind a corrugated screen of shimmering translucent plastic panels that run perpendicular to the main dividing wall.

Photographs: Paul Warchol

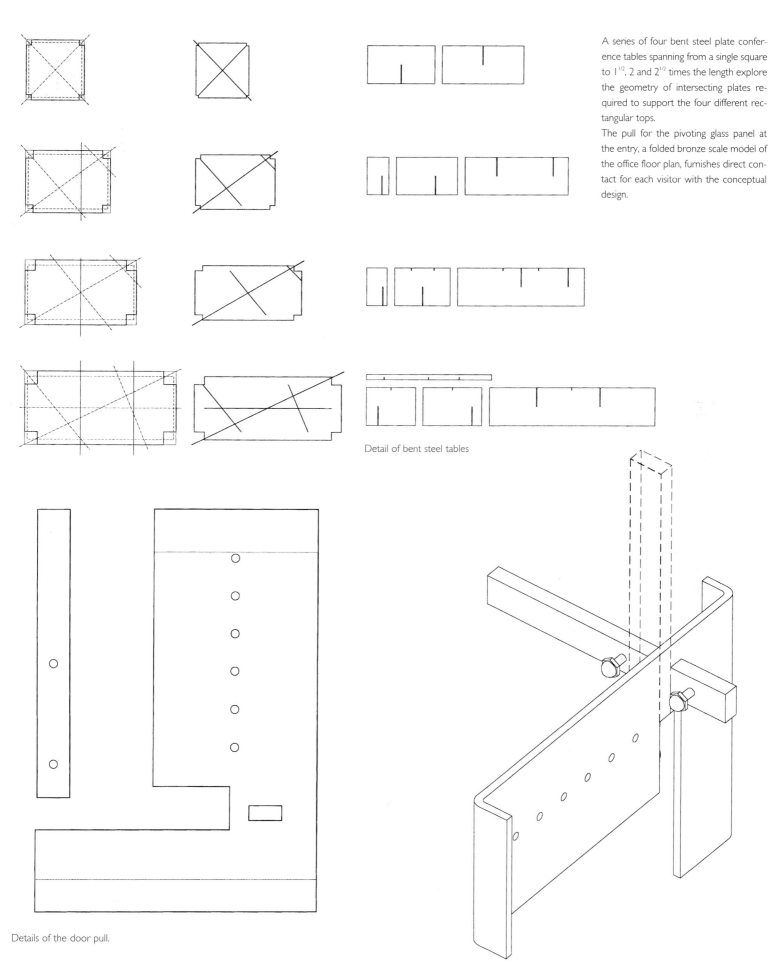

A series of four bent steel plate conference tables spanning from a single square to $1^{1/2}$, 2 and $2^{1/2}$ times the length explore the geometry of intersecting plates required to support the four different rectangular tops.

The pull for the pivoting glass panel at the entry, a folded bronze scale model of the office floor plan, furnishes direct contact for each visitor with the conceptual design.

Detail of bent steel tables

Details of the door pull.

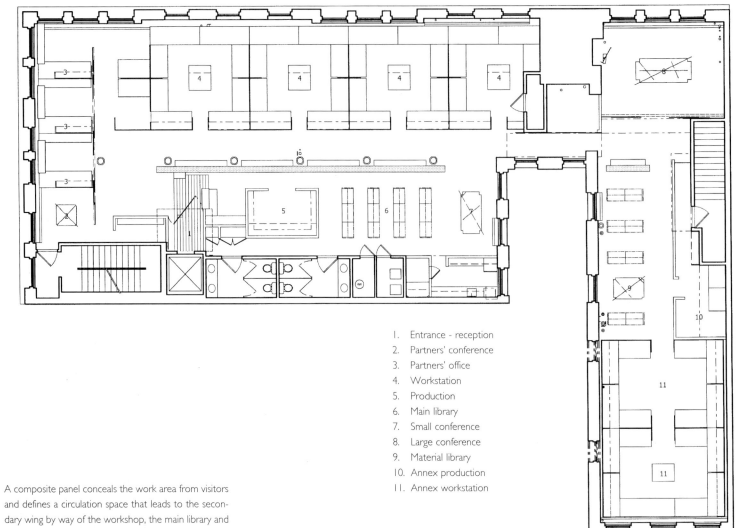

1. Entrance - reception
2. Partners' conference
3. Partners' office
4. Workstation
5. Production
6. Main library
7. Small conference
8. Large conference
9. Material library
10. Annex production
11. Annex workstation

A composite panel conceals the work area from visitors and defines a circulation space that leads to the secondary wing by way of the workshop, the main library and the small meeting room.

de Architectengroep
(Bjarne Mastenbroek)
Conversion and Extension of a Culture and Education Center

Den Helder, The Netherlands

In spite of its apparent complexity, this is a surprisingly low-cost rehabilitation and extension of a former school building. Like a parasite, the new wing takes hold of the existing structure: the glazed facade of the new wing also wraps around and enfolds the long nave of the old school house and the bathroom units.

The extension is therefore extremely efficient in terms of overall floor space. All functions requiring different ceiling heights are situated on the first floor of the extension.

In this cold northern clime with short winter days, the maximizing of natural light was a central preoccupation in the extension. The entire ground floor elevation along the front of the building is glazed. Both floors of the two sides of the new wing are also completely glazed; while a sense of uniformity between the new and the old is achieved by continuing this glazing around the existing structure.

A series of smaller windows and skylights have been haphazardly punched into the various sloping planes of the new roof, creating interesting lighting effects in the interior during the day. Externally, these small windows serve to break up an otherwise imposing and bulky roof.

The new wing is covered by a steel construction with a traditional wood and bitumen roof, which is in turn covered by a skin of Western Red Cedar planking. This outer skin serves as a sort of elegant camouflage, while also shielding the front of the building from too much direct sunlight.

Photographs: Christian Richters

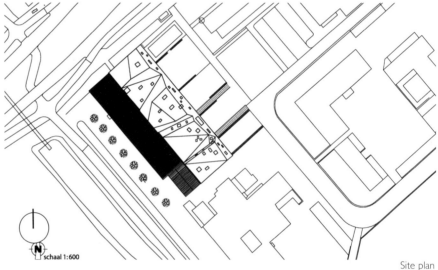

N schaal 1:600

Site plan

The long nave and bathroom units of the original school house are the anchor around which the new project is wrapped. Three of the facades have been almost entirely glazed in order to bring in the maximum amount of natural light - a central theme in this project due to the cold northern clime.

Upper floor plan

 1. Rehearsal (piano room)
 2. Foyer
 3. Conference/performance hall
 4. Photography studio
 5. Music room
 6. Soundproofed music room
    (drum room)
 7. Drawing room
 8. Storage room
 9. Meeting room
10. Sound control room
11. Photo lab
12. Sculpture room
13. Pantry
14. Dance room
15. Assembly room
16. Heating
17. Audio-video
18. Wardrobe
19. Administration
20. Consulting
21. Coordination
22. Media archive
23. Sculpture storage
24. Bathroom

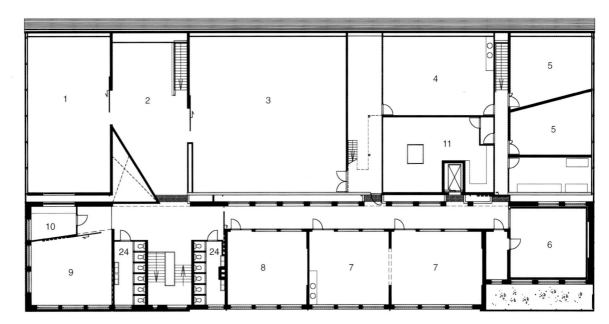

Ground floor plan

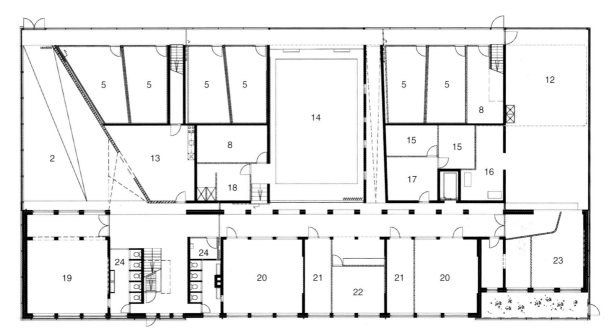

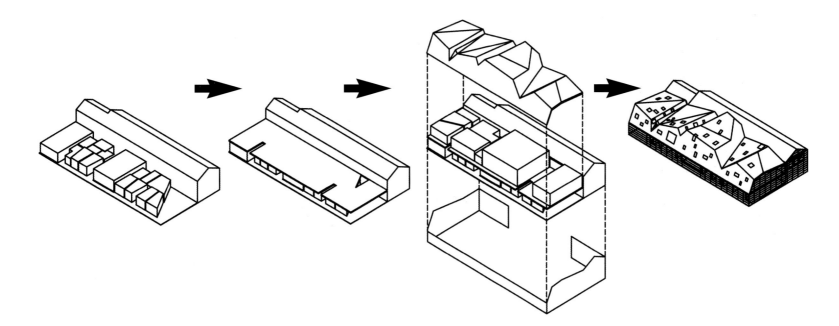

The new wing is covered by a steel construction with a traditional wood and bitumen roof, on top of which is a skin of Western Red Cedar wood planking, which serves as camouflage and heat insulation.

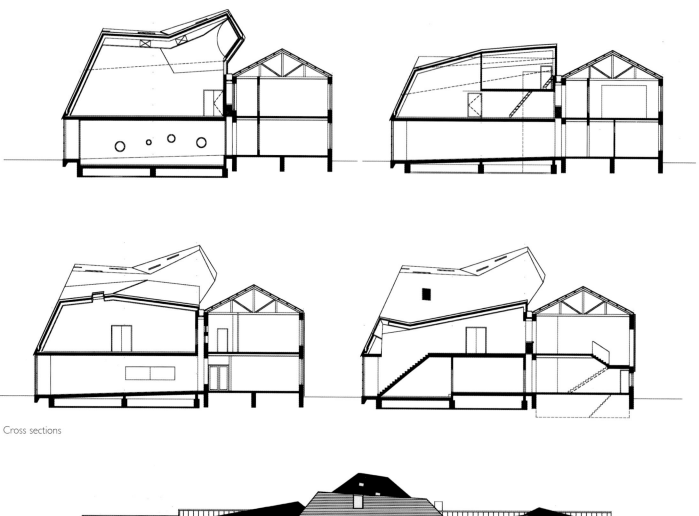

Cross sections

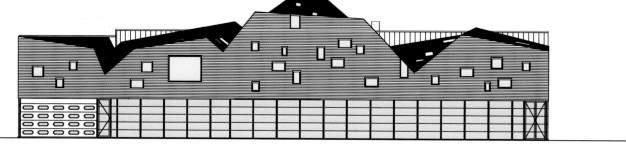

Main facade

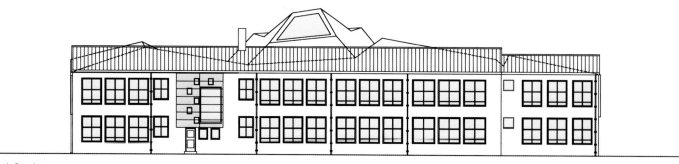

Back facade

Longitudinal sections

Rataplan
Bürombau Vienna Paint

Vienna, Austria

An industrial workshop dating from 1899 was converted to provide the offices of a digital company. The commission was to contain workstations, a computer room, scanner room, film processing equipment, etc. all of which would be cross-linked. Each of these fields also had to correspond to different requirements of acoustics, lighting and climate.

It was a very important starting point of the architectural concept to create no cells but to conserve the originally open space and generate views.

There are offices on the first floor, an exhibition space on the ground floor and a coffee house in an annex. The entrance area is marked by a horizontal style plate, which acts as a canopy, and a vertical one that leads to the depth of the space. A new staircase leads to the upper level where the offices are located. This staircase is formulated as an upright element linking the two floors. The existing elevator was partly exposed by removing a wall and part of the ceiling. The windows have been enlarged and transformed; now they give views of the industrial chimney and allow it to function inside the space.

In the upper floor the horizontal composition remains, by means of three freestanding, articulated shelf elements. As with everything new in the building, these elements are set at an angle of 11° to the existing walls from the backbone of the space and accentuate the perspective. All abutments to existing walls and the roof are in glass to maintain the sense of spatial continuity.

On account of the different requirements, it had to be possible to close off the individual areas. Between the closed areas are the "work bays" of the zones without special acoustic and climatic requirements. The office in the middle of the space represents the 'market place' where clients are received.

Photographs: Markus Tomaselli

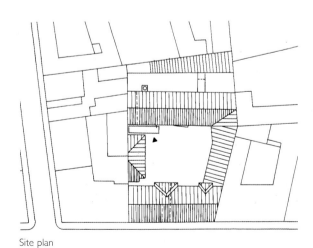

Site plan

This industrial workshop was built in 1899; its interior was totally remodelled in order to house new offices.

Here, detailed views of the staircase leading to the upper level, which has been maintained as a single, indivisible space.

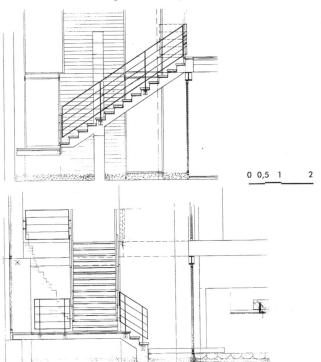

0 0,5 1   2

Construction detail of the staircase.

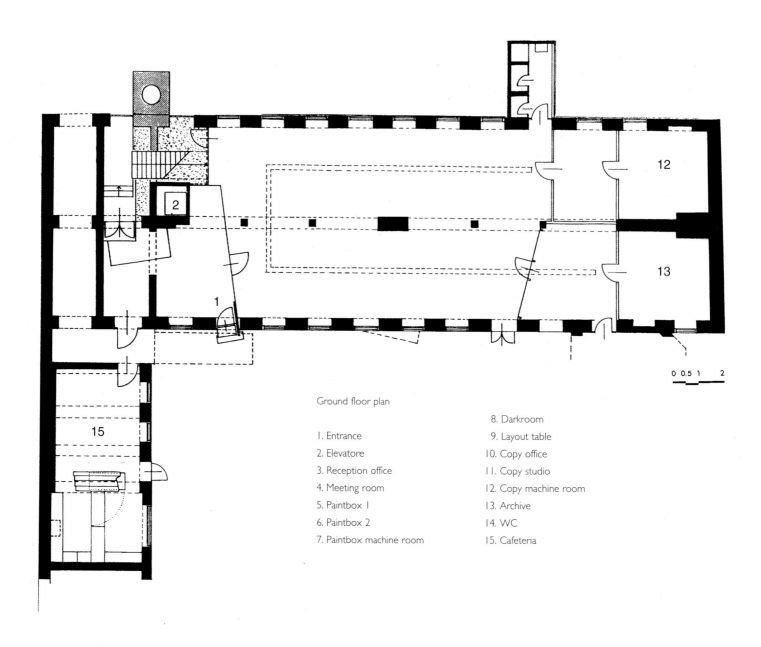

Ground floor plan

1. Entrance
2. Elevatore
3. Reception office
4. Meeting room
5. Paintbox 1
6. Paintbox 2
7. Paintbox machine room

8. Darkroom
9. Layout table
10. Copy office
11. Copy studio
12. Copy machine room
13. Archive
14. WC
15. Cafeteria

0 0.5 1 2

First floor plan

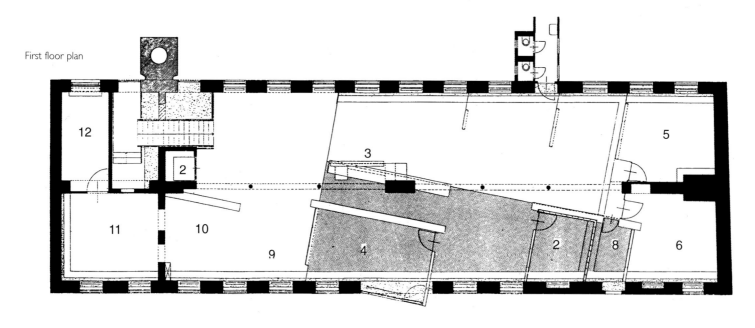

Three self-supporting shelving units placed at an angle of 11° against the walls of the building articulate the space and accentuate the effect of perspective. The elements of glass and perforated metal plate appear either transparent or opaque, depending on the lighting conditions.

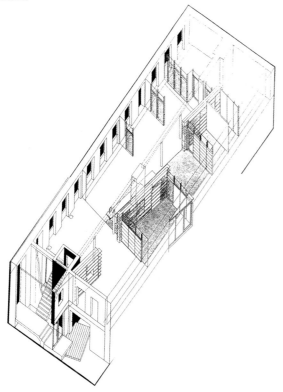

Eduardo Souto de Moura
& Humberto Vieria
Inn at Santa Maria do Bouro

Amares, Portugal

This project aims to adapt, or rather to make use of the stones available to create something new. This is a new building, in which various voices and functions (some already registered, others still to be constructed) intervene; it is not reconstruction of the building in its original form.

For this project, the ruins are more important than the Convent. It is they that are open and manipulable, just as the building was during its history. This attitude is not meant to express or represent an exceptional case justifying some original manifesto, but rather to abide by a rule of architecture, more or less unchanging through time.

During the design process, lucidity between the form and the program was the desired aim. Faced with two possible approaches, the architects chose to reject the pure and simple consolidation of the ruin for the sake of contemplation, opting instead for the introduction of new materials, uses, forms and functions *entre les choses*, as Corbusier said. The picturesque is a question of fate, not part of a projection or program.

Photographs: Duccio Malagamba

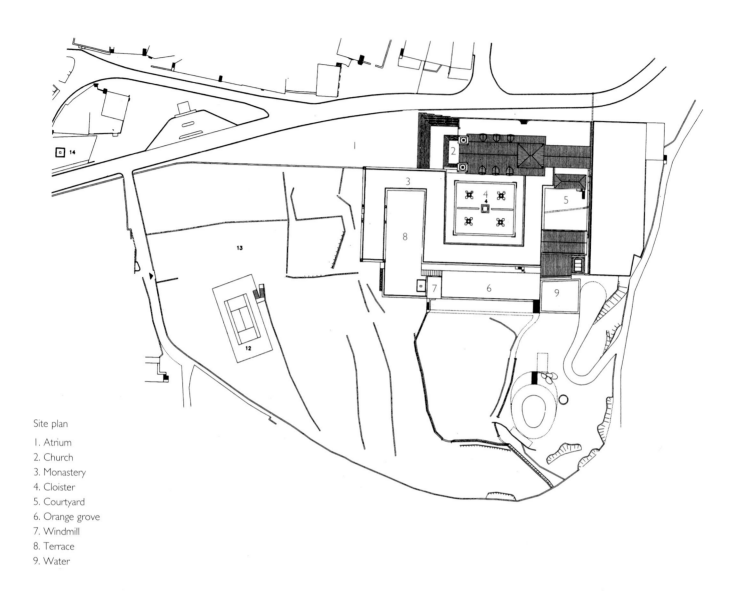

Site plan

1. Atrium
2. Church
3. Monastery
4. Cloister
5. Courtyard
6. Orange grove
7. Windmill
8. Terrace
9. Water

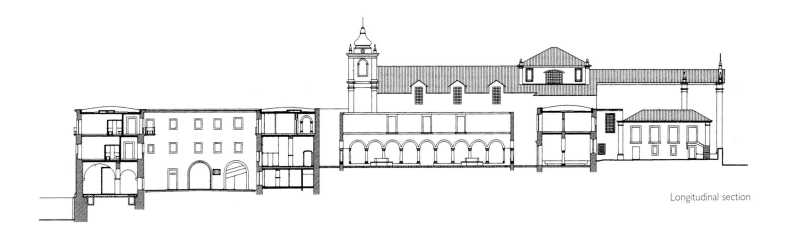

Longitudinal section

The most important value of this project was the meticulous work carried out over several years by Souto de Moura, in which he studied and reinterpreted the ruins of the old monastery to transform it into an inn. The south facade, with the garden and pool, a long volume housing the bedrooms of the hotel staff, and the east wing of this facade into which the restaurant is inserted are the newest elements of the restoration, which was recreated under the suggestions inspired by the shapes of the ruins themselves.

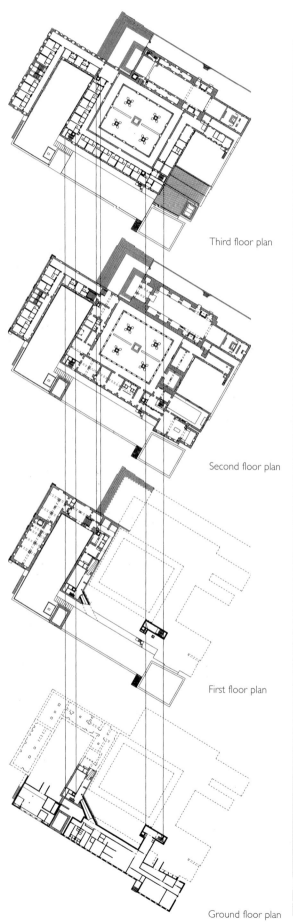

Third floor plan

Second floor plan

First floor plan

Ground floor plan

Inside the building, the space was organized so as to make it as transparent and manipulable as possible. The architect thus brings ample lighting to all corners of the old convent.

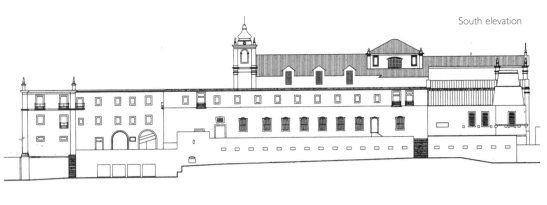

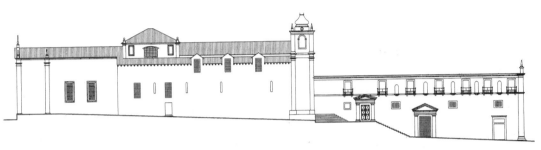

43

Both in the design of the new spaces and in the consolidation of the existing ones, one of the most prominent aspects is the introduction of new materials and forms that do not belong to the original structure of the building.

Vertical section

0   0.2   0.4   0.8

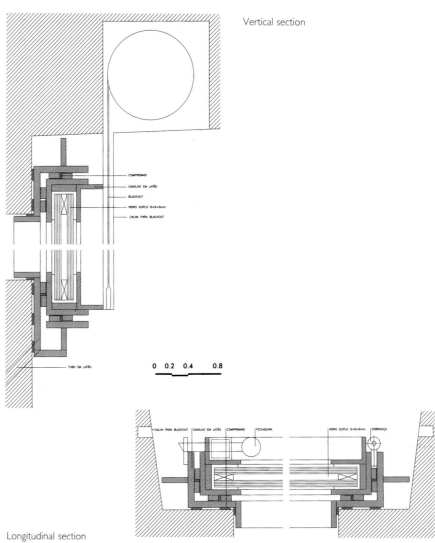

Longitudinal section

44

Vincen Cornu & Benoit Crepet
Museum of Ambulant Theater

Artenay, France

After the municipality of the French village of Artenay inherited some years ago the stage property of a troupe of itinerant players, it was decided to create a museum of travelling theater as the centerpiece of a whole area's renovation. Apart form the museum, the program was rounded out by a local archaeological exhibition, reserves and workshops for the museum of travelling theater, plus a documentation center and a small public library.

The architects saw this set of ordinary rural buildings as a landscape, bringing their diversity into a coherent whole and restoring an old itinerary around a new communal facility.

In order to conserve the balance of the place while asserting its new vocation, the architects magnified the walls, which they saw as vital to its identity, and manifested the presence of the new interior facilities by way of the openings.

Windows and doors were redistributed on the existing facades, restored with a careful eye for traditional stonework details. Woodwork and doorways were treated as noble elements and built by local craftsmen. The new wing, designed to close the yard of the Paradis farm that backs onto the mall, shows the same concern or unity and dialogue with neighboring forms.

Permanent exhibition rooms are housed in what was once the main barn of the farm, the structure of which was laid bare and the render renewed. Linked by ramps and a footbridge, they compose a complex itinerary distributed over two levels on either side of the full-height central volume that structures the whole.

Materials suggest refined rustic taste: terra cotta floors, solid woodwork, and render painted white to distinguish restored walls from partitions. The original door was enlarged and rebuilt with particular care. When it is open wide, the central volume is opened to nearly six meters, transforming the space into a stage for spectators gathered in the courtyard.

Photographs: Jean Marie Mothiers, Benoit Crepet

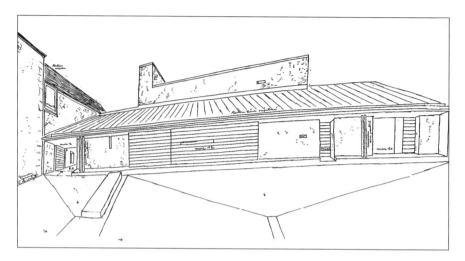

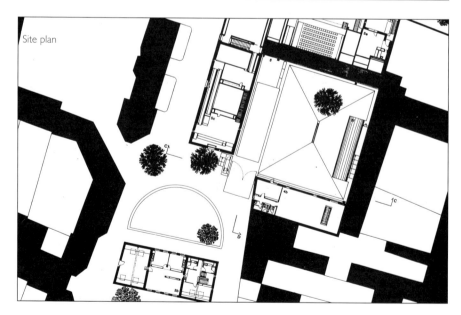
Site plan

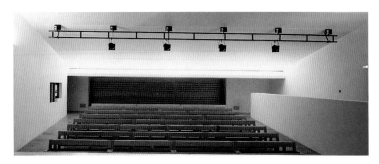

The architectural vocabulary of the rehabilitation is attentive to the character of the place and to a knowledge of the local architecture.

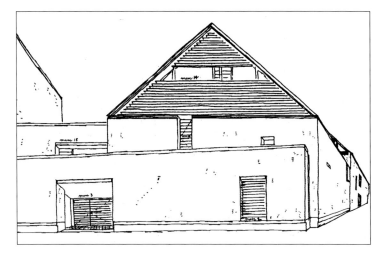

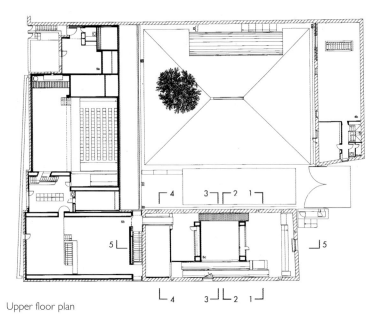

Upper floor plan

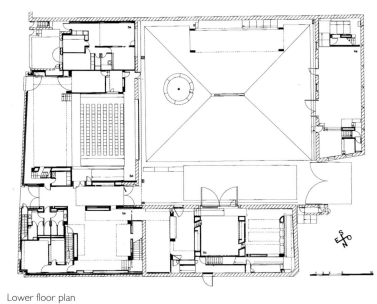

Lower floor plan

The program consists of the rehabilitation of a set of buildings of agricultural origin. The vocabulary, based on the dialogue between white cloth and woodwork, seeks to unify the buildings.

The museum spaces were organized in an orderly sequence in accordance with the rhythm of life of a theatrical company at various times in history. The assembly shows elements referring to the arrival of the actors at a village, various aspects of the performance and their departure.

Section 1-1

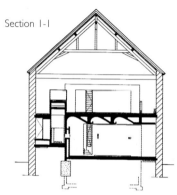

Section 2-2

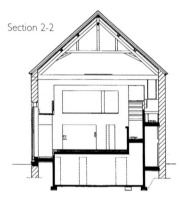

Section 3-3

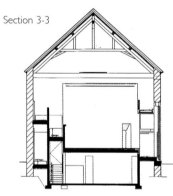

Section 4-4

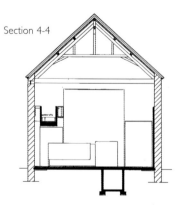

Section 5-5

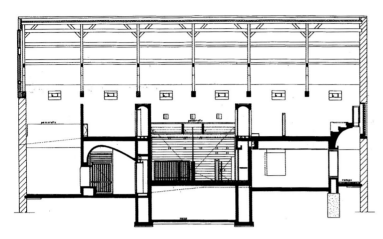

Gunter Domenig
Landesusstellung Kärnten

Hüttenberg, Austria

On the remains of an old steelworks that had been abandoned at the beginning of the century, the Austrian architect Gunter Domenig made a reinterpretation of the large architecture of the factory, melted with new forms, in order to create a modern conference and exhibition center.

The new work persistently uses steel, the material that used to be produced here, to reveal aspects of process, building and culture. The accommodation schedule included a large lecture hall and about 10 smaller meeting rooms as well as exhibition space. Despite cuts made to the original design for budget reasons, Domenig´s most important measures are largely visible: the new multipurpose hall somehow seems to float over the street, an organic, windowless volume under a metal skin. Over it, also floating: is an awe-inspiring horizontal steel and glass construction - the built, architectural association with a gallery that runs through the whole plant like a high speed train and ends stunted and incomplete, as if it were to be continued at some time in the future. It ties the different buildings together.

The glazed part of the gallery was initially considerably longer. However, it is still very impressive to view the old walls from this new perspective, or to look down from one of the two small balconies into the machine house.

Domenig closed the openings of the ruin very simply, with single glazing; and he did not replace the missing gable roof but rather covered it with a flat roof that leaves the gables standing free. Of the glass roof of the initial project, only two narrow skylights remain. Although this compromised the beautiful illumination intended by the architect, it does not affect the substance.

Photographs: Gerald Zugmann

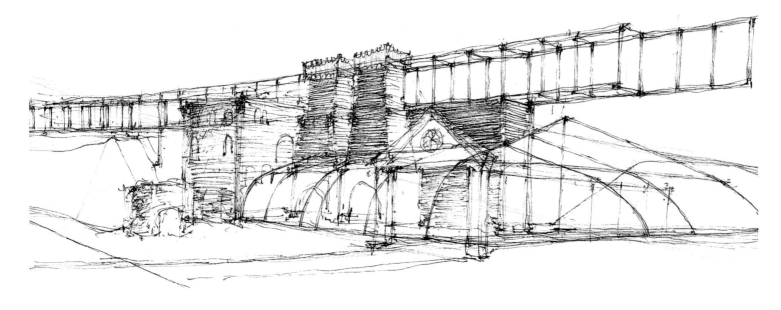

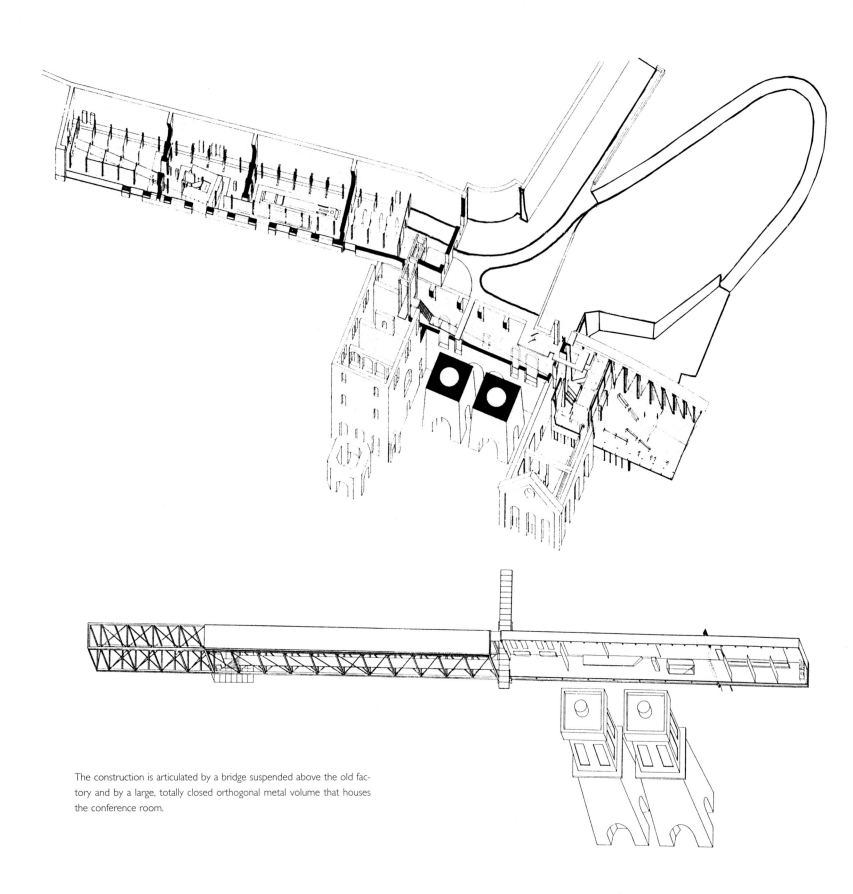

The construction is articulated by a bridge suspended above the old factory and by a large, totally closed orthogonal metal volume that houses the conference room.

Ground floor plan

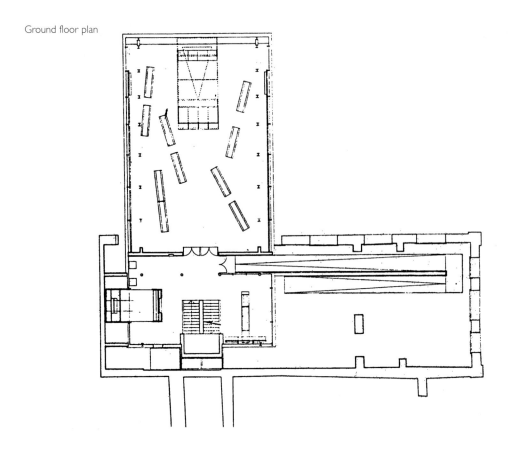

Upper floor plan

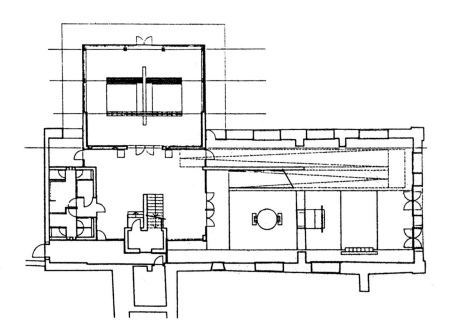

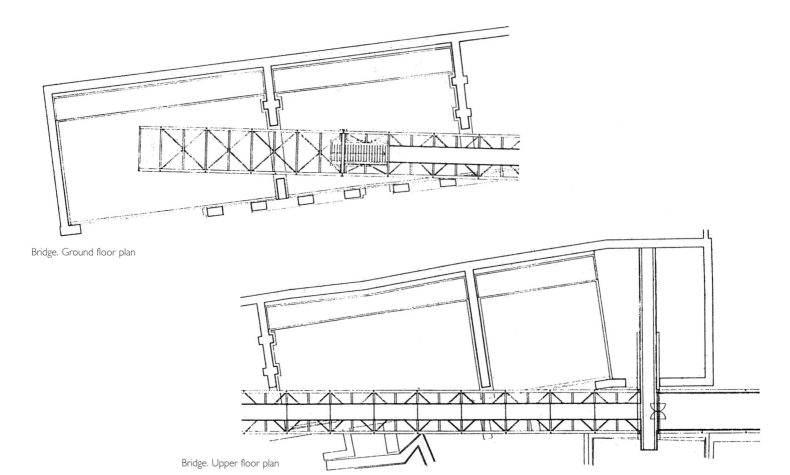

Bridge. Ground floor plan

Bridge. Upper floor plan

The new intervention contrasts with the old steelworks in an energetic dialogue between the industrial aesthetics of the new addition and the sober volumetrics of the existing building. From the bridge, the visitor has a striking view of the archaeological remains of the factory.

Ramón Esteve
Dwelling Between Party Walls

Ontinyent, Spain

This project consisted of rehabilitating a terraced single-family dwelling located in an area that was originally developed outside the old town center, but at the beginning of the 20th century it was the only remaining nucleus of the town.

The original layout of the house followed the lines of early urban dwellings for country people; a large front door giving direct access from the street, a narrow staircase, small and little-used inner courtyards, and a back garden that was used as a vegetable garden or to store farm implements. The restoration respected this initial layout and did not distort the essential structure of the dwelling. Practically all the secondary elements of the house were demolished, the walls were stripped, the floors were taken up, and all the floor slabs were demolished and reconstructed with the old wooden beams that were salvaged. The floor bricks were specially made of hand-made terracotta for the house. The bricks used to clad the courtyards and the floor tiles are made of hand-made terracotta, the doors and furniture of solid Iroko wood, and the external door and window frames are of zinc-plated steel.

The design is articulated on three floors with the following distribution: access from the street to the ground floor, housing the entrance hall, the garage, a cellar, an office, a toilet, the straight main staircase, and another open staircase that communicates the light well with the rear garden. The main staircase leads to the dining room, and from here a single step down gives access to the living room. The kitchen receives light from the garden through a french window that is also the exit to the garden. The garden is about the same size as the ground plan of the house, and its main feature is a lemon tree. From here, an external staircase leads to the second floor. A hall on this floor opens the way to two single bedrooms, two bathrooms, a dressing room and the main bedroom with a balcony overlooking the garden.

Photographs: Ramón Esteve

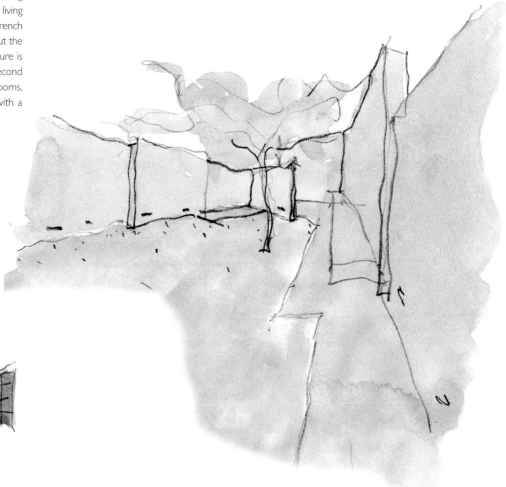

The essence of the projects lies in the harmony of the materials with the lighting, which creates a masterful interplay of interesting tones and textures in the whole house.

0   5   10

Ground floor

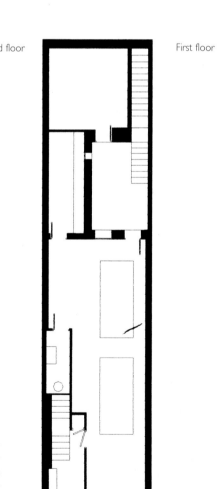

First floor

Second floor

The house has a longitudinal floor plan, with the spaces aligned according to a logical rectilinear structure that crosses the whole building from the bustle of the street to the tranquillity of the garden.

In this scheme the traditional building techniques, rooted in Mediterranean culture, adapt perfectly to the contemporary dwelling concept. The final result is an ordered combination of light, clay, iron and wood.

Luigi Ferrario
Home Studio for a
Graphic Designer

Bergamo, Italy

In a space with an unusual arrangement, in a traditional old Lombard three-story house, Luigi Ferrario experiences, through a rarefied language, his own vocation for dialogue with old, sedimented forms and materials.

The architectural shell is characterized by a vertical volume of just 7 m², used for the bathroom, the staircase, the kitchen below and the sofabed in front of the fireplace, connected above to a work area distributed horizontally in relation to the house.

The domestic space features an entrance at the middle floor, a covered vaulted staircase connecting a courtyard that does not look onto the studio and a single living room on the top floor.

The entire design centers aroun the connection between the entrance and the bathroom, distributed in a single, narrow and vertical environment connected to the attic.

A small opening that was created when, some time ago, the end of the barrel vault that partially covered the original stair collapsed, creates a "natural" cavity of only 2 m² providing access to the next floor.

The available space has been transformed without disrupting the features of the original structure: the vault, the floor above it in terracotta with inclined steps, and the stone masonry.

The introduction of an original structure in iron, glass and wood succesfully modifies the space and connects the two floors: through careful additions and minimal subtractions it has been possible to cater for all the functions necessary for domestic life without having to subdivide the available area, so as to obtain space for the indispensable kitchen and bathroom.

Photographs: Alberto Piovano

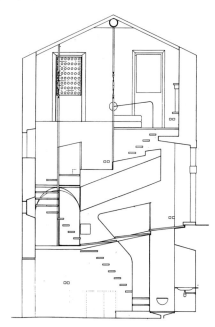

Cross section

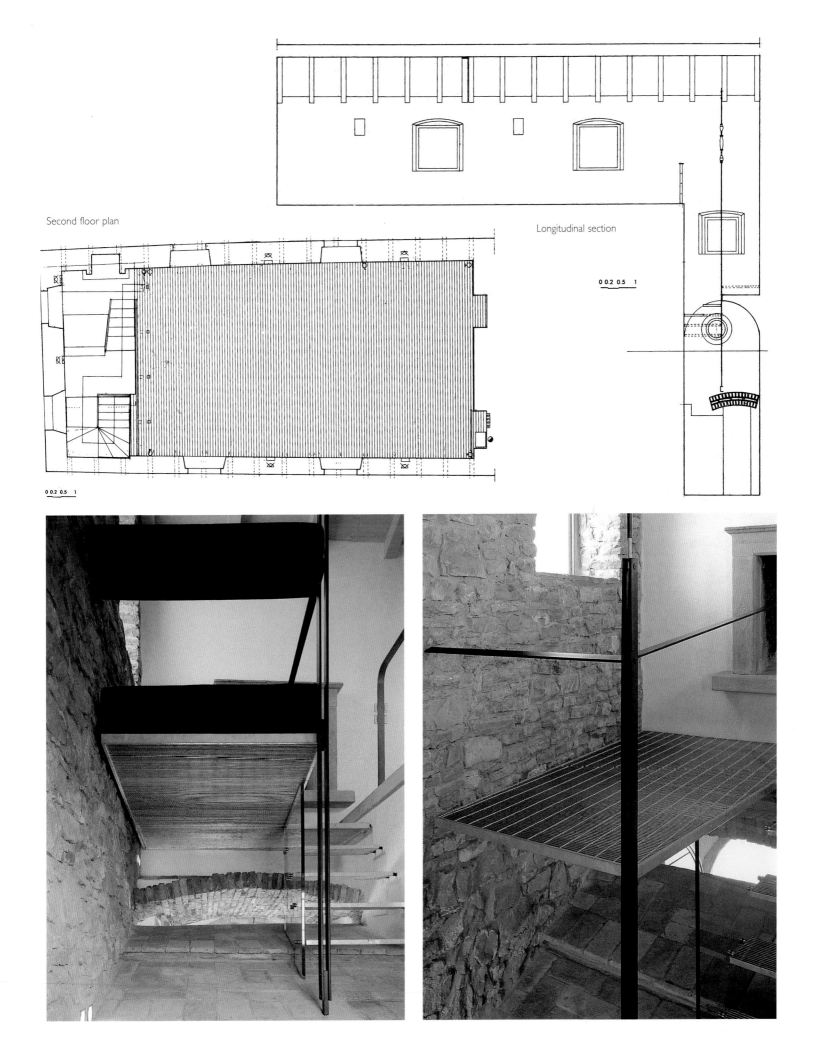

Second floor plan

Longitudinal section

0 0.2 0.5 1

0 0.2 0.5 1

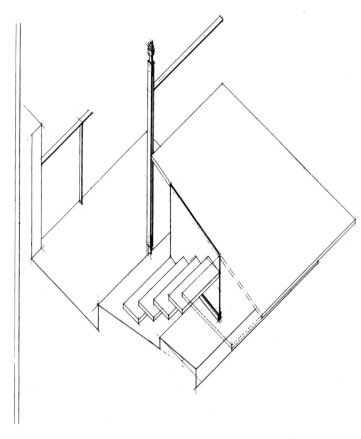

The bathroom and the small kitchen located on the ground floor are designed to avoid excessive subdivision of the space. The bathroom has a satin-finish sliding glass door.

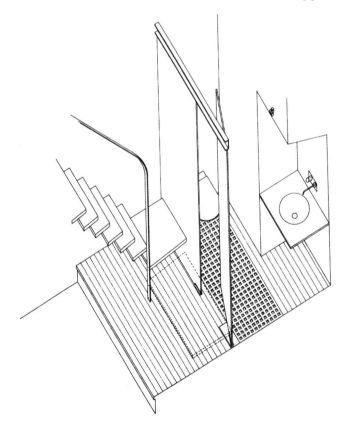

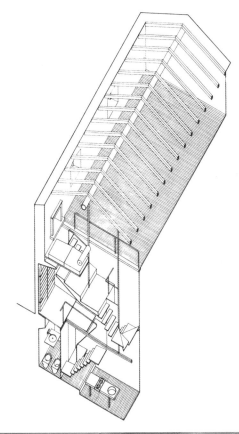

Architekturbüro Gasparin & Meier
Badehaus Ebenberger

Siflitz, Austria

Located 1000 meters above sea level, this old two-story stone farmhouse was a damp place without running water that was urgently in need of restoration. The absence of running water aroused the desire to create a special area for the bathrooms without spoiling the open plan of the existing structure.

Access to the building is through a large, striking entrance located in the center of a stone wall. This door gives access to two large rooms located on both sides of this entrance that are used as bedrooms.

In the essential restructuring, the tarmac road that goes down the mountain had to be moved because it was located too near the building. The foundations were left free to protect the walls from damp. The wooden bathroom annex located to the west is now the sunniest area in this solid stone house.

This room, which was formerly used for smoking bacon, now houses the bathroom with its washbasin, bathtub and other complements. It is a spectacular space which also gives access to a corridor measuring 8 meters in length by 1.5 meters in width. It is used as a luxurious sauna and as a bridge that communicates with the old area of the building. One of the most striking aspects of this scheme is the importance given to the bathroom, which is further enhanced by the large sliding window that offers excellent views of the valley. Another of the outstanding features is the use of stone cut from the higher part of the mountain for the pavement and part of the walls.

Under the old stone wall, which had to be demolished and reconstructed, a new basement was created taking advantage of the old staircase. It receives light through the glass at the bottom of the bridge.

Photographs: Margherita Spiluttini

A new wooden annex in the west area serves as a bathroom and sauna for an old farmhouse. This intervention also improves the natural lighting of the house.

Besides the splendid views from the bathroom, the contrast between the larch wood and the old stone creates a suitable framework for contemplation and relaxation.

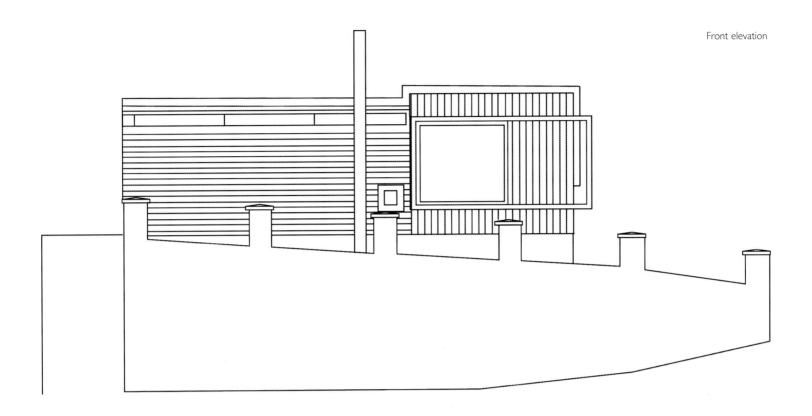

Alzado lateral

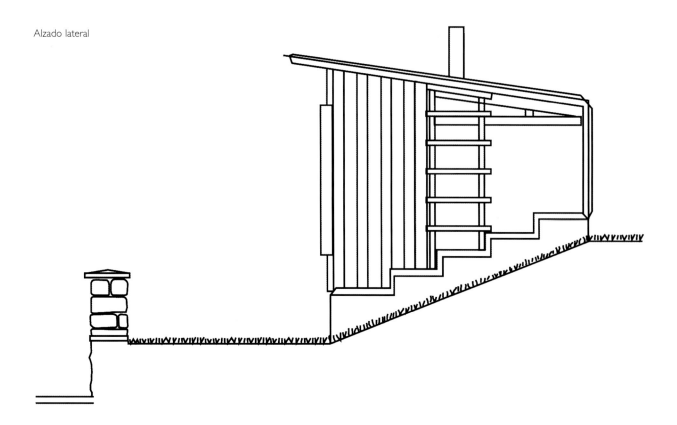

Crone Nation Architects
Establishment Hotel

Sydney, Australia

In 1996, a fire destroyed almost half of this 100-year-old structure. In its intact state, the George Patterson building was a rare and fine example of 19th century commercial architecture and was unusual in combining both retail and warehouse functions. The George Street elevation made use of high-quality materials (stone, bronze and brick) and features finely detailed windows, arched bays to the street with a rusticated sandstone base and fluted pilasters at the upper levels.

After the fire, the rear, former warehouse, section remained as a four-story building, and the front section, including the cast iron support columns, survived in an extensively damaged state. The tower survived in close to its original form prior to the fire. The renovation strategy sought to preserve as much of the original as possible. New work, while closely following the configuration and material for the earlier work, does not copy it exactly. This allows the observer to understand what is original and what is not.

Externally, materials generally follow or approximate the original work with minor variations. The junction between new and old is delineated along the north, east and south elevations by a band of red bricks. As much as possible of the smoke staining and other evidence of the fire has been retained on the various elevations. The tower is retained as a semi-ruined structure.

Internally, original brick and plaster surfaces have been left exposed where they are sound. The iron columns in the George St building have been left exposed with those in the Garden Bar retaining evidence of the fire and subsequent exposure.

Ceilings in the Tank Stream Building were originally unlined and they remain so. The joist structure with beams, herringbone strutting and the underside of the flooring is visible within hotel rooms and in public spaces. Services have been confined to corridors and lesser spaces, where they are set below the structure.

Photographs: Phillip Hayson

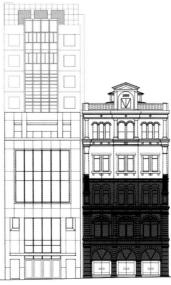

West elevation

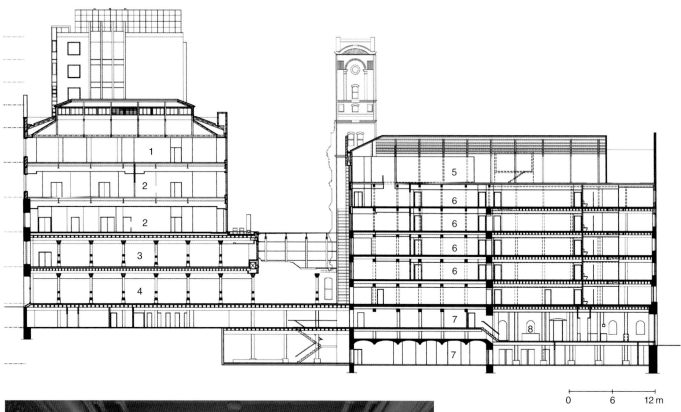

Longitudinal section

1. Private lounge
2. Main function area
3. Restaurant
4. Main bar
5. Floor
6. Hotel floor
7. Night club
8. Hotel lobby

0 6 12 m

South elevation

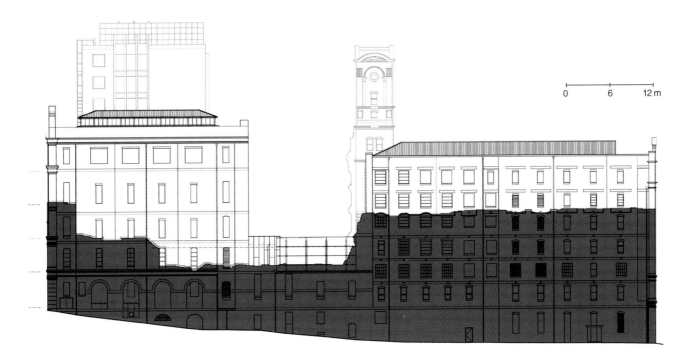

Almost all existing windows, both steel and timber, have been conserved. The new fire and acoustic strategy required that all of the timber floors be covered with a topping slab. Voids in the existing floor structure have been expressed by the use of different joist configurations between the infill panel and existing structure.

Almost all of the timber and iron columns have been left exposed, with those in the Garden Bar retaining evidence of the fire. Ceilings lined with timber and pressed metal have been repaired. The joist structure with beams, herringbone strutting and the underside of the flooring is visible in hotel rooms and in public spaces.

2nd and 4th floor plans of hotel

1. Kitchen
2. Main function area
3. Hotel

Jean-Paul Philippon
Museum of Art and Industry

Roubaix, France

The transformation of these old baths into a museum is an excellent example of the possibilities offered by modern architecture for adapting a large space to a use other than its original.

This early 20ᵗʰ century building once served the cult of the body as well as the spirit. While the saunas, baths and pools were meant for bodily hygeine, the volumes, light and spatial distribution are reminiscent of convents. Thus, the artwork now put on display here loses none of its value; rather, it can be appreciated with a more privileged clarity.

Work done to the structure was based on a concern for how the museum is viewed and for the composition of the walkways. The addition of a new wing, which finalizes the paintings segment, and of a room for temporary exhibits was done by geometric inference of the existing construction. The former lies parallel to the pool and effectively concludes the visit to the Belles Artes section, located in the old baths. The latter is an extension of the space between the pool and the industrial facade of l'Espérance street, thereby defining the space intended for temporary exhibits and the auditorium.

The lobby sits below a long steel coat within a transparent volume that wraps the old structures, the perimeter wall and the cafeteria. This serves as a connection to all the other areas and also lets light and views pass over the upper bridge housing the teaching workshops. Horizontally, the eye is drawn to the farthest views of the pool, garden or temporary exhibit; vertically, following the path of the service elevator, one's attention is led upward. All paths converge at the entrance to the museum, in the "compass room", which is the former workout room in which vestiges of the swimming pool can still be seen.

A central strip of the old swimming pool, whose original mosaic edgings peek out along the periphery, still holds water and is delimited by wood platforms supporting the sculpture exhibit. Refracted from the water's surface, light glimmers on the glass cases and sculptures, which seem to tremble in the shimmering light.

With highly worn steel due to the humid, chlorinated air, the original vaulted ceiling was in danger of collapse. The massive arches of the roof and intrados were freed and then strengthened with restored armature. A new exterior rustproof roof was installed; while perforated plasterboard has been suspended from the interior face, providing acoustic padding and ventilation.

Photographs: Arnaud Loubry

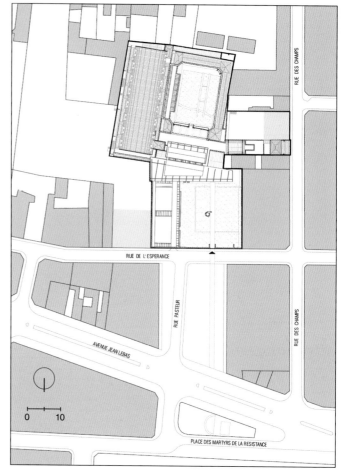

Site plan

 Plot line
 Converted existing building
Extension

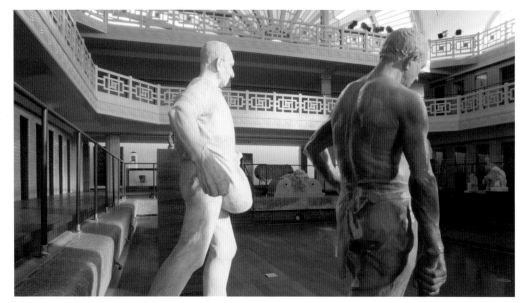

Being highly deteriorated, the tympana at both ends were restored with new glass brought in from all over the world due to the difficulty of finding stained glass from the original era. This glass is mimicked on the exterior by another layer, contributing to better conservation and temperature control.

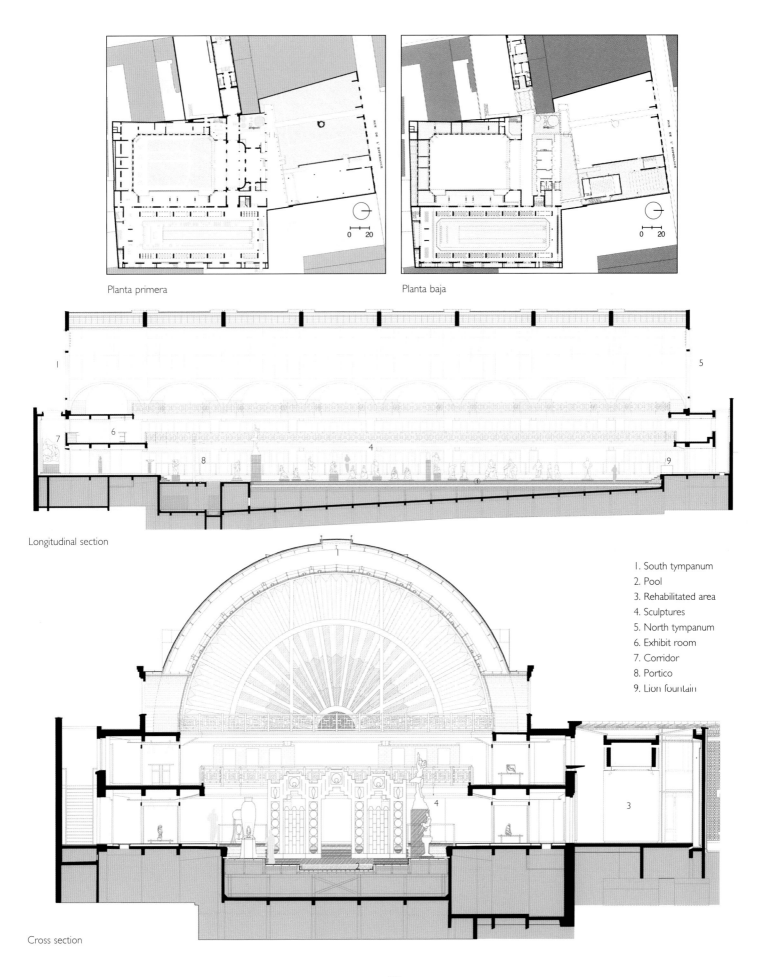

Planta primera

Planta baja

Longitudinal section

Cross section

1. South tympanum
2. Pool
3. Rehabilitated area
4. Sculptures
5. North tympanum
6. Exhibit room
7. Corridor
8. Portico
9. Lion fountain

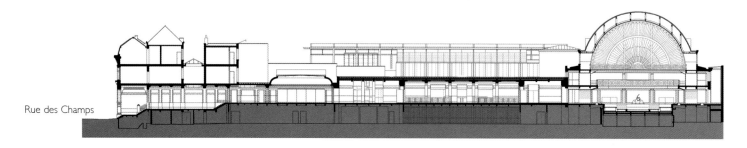

Rue des Champs

Secciones del conjunto

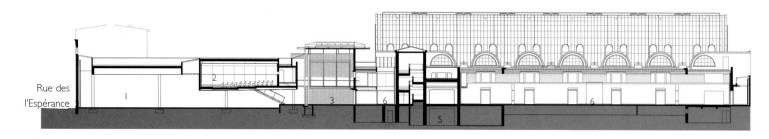

Rue des l'Espérance

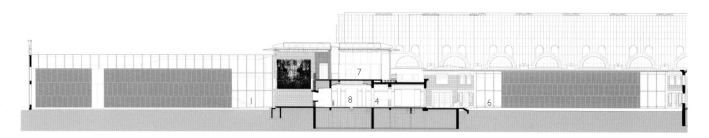

South sections

1. Temporary exhibit room
2. Auditorium
3. Hall
4. Meeting room
5. Storage
6. Rehabilitated area
7. Learning activities room
8. Restaurant

Exit routes from the restored gallery housing the pool lead either to the left, toward the two halls dedicated to the contemporary artwork of the Roubaix school, or to the right, toward the cafeteria, shop or garden.

Klaus Block Architekt
Church of St. Mary
Conversion and Library

Müncheberg, Germany

The 13th century Church of St. Mary is the city's most emblematic and widely visible landmark. Damage done on the structure during WWII left the building in ruins –without a roof or vault– until 1992, when renovation work began. Partly for financial reasons, the municipal library was moved into the nave of the church; it has been conceived as a free-standing volume within, yet apart from, the church.

The interior building strongly suggests a ship motif. The broad side of the interior building lists to the east by the same measure as the slope of the top of the new construction, the end of which is level with the height of the central vault, which in turn serves to divide the library and choir.

A new elevator tower, connected by gangways to the library, acts as a counterweight to the curving of the new volume. It has a free-standing steel frame with no structural connection to the interior building and is clad in perforated sheet metal.

The interior building is climatically and acoustically autonomous, thereby creating a flexible space which may be used for seminars, conferences and cultural events. The wall of the storage room facing the interior hall can be opened similar to a market stand and, when open, serves as a canopy above a small stage area that can be erected.

The primary structural element is an extremely minimized steel frame not connected in any way to the existing historic building. It is stiffened with cross-bracing within the bookshelves and on the ground floor with concrete slabs which in turn function as utility room walls. The floor slabs are located flush between the beams and consist of 93 mm reinforced concrete with an integrated floor heating system and a body coat.

The three sides facing the interior church space are clad with horizontal ash slats which run perpendicular to the arched steel columns.

Design Team: Susanne Günther, Heike Simon, Siegfried Casteleyn
Landscape planning: Gabriele Schultheiss

Photographs: Ulrich Schwarz

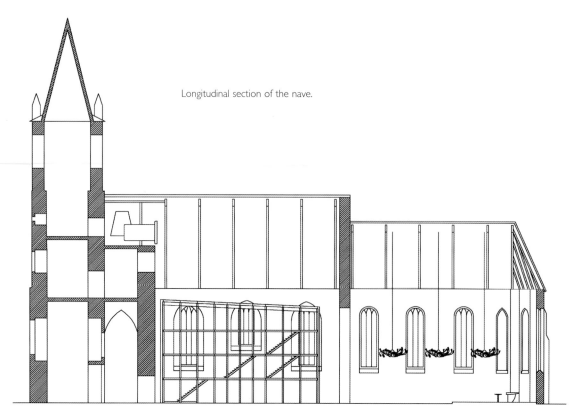

Longitudinal section of the nave.

Floor plan of library

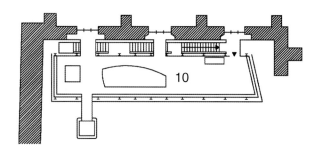

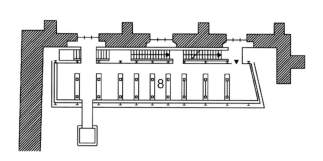

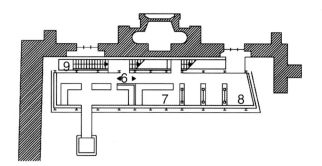

1. Wardrobe
2. Kitchen
3. Seating storage
4. Bathrooms
5. Elevator
6. Library entrance
7. Loan desk
8. Book stacks
9. Main staircase
10. Conference room

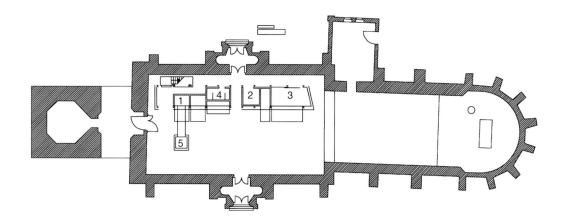

The roof of the 13th century church was severely damaged in WWII and had to be replaced. The free-standing church tower served as inspiration for the new library's independent elevator shaft inside the building. The exterior landscaping was also done in conjunction with the project for the library.

Four floors lie within a narrow volume running along the "ship's" outer wall, allowing a maximum of natural illumination and ventilation while covering a minimum of floor area. The church has been fitted with new window panes, but all the weight has been transferred to a new steel frame instead of the old masonry.

Construction detail of window.

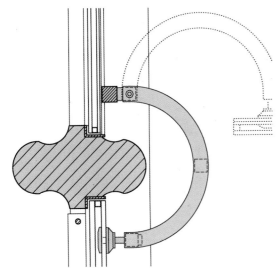

Detail of folding door

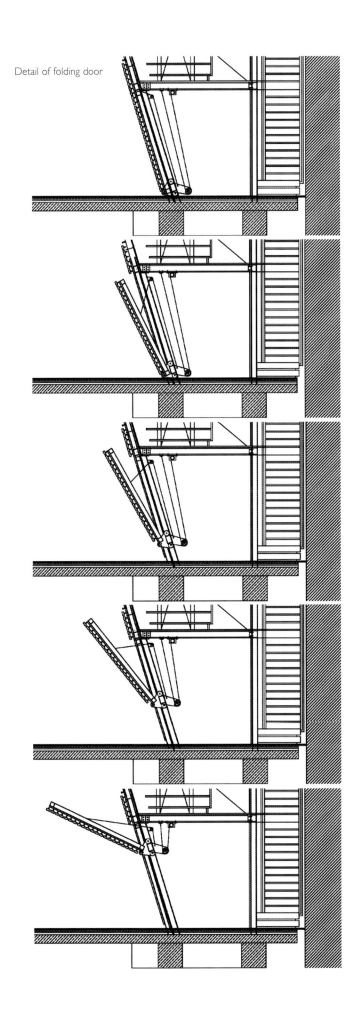

Library section

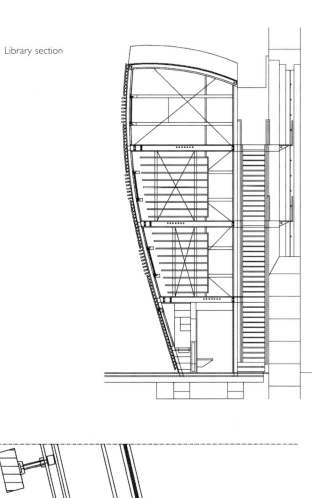

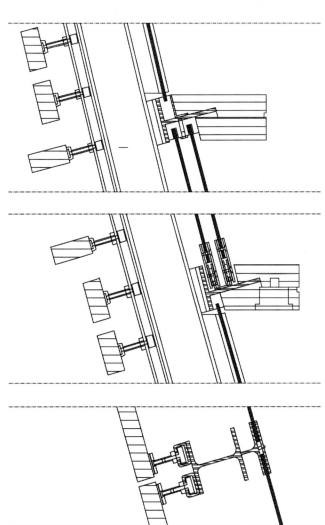

Roberto Luna / Arata Isozaki
CaixaForum

Barcelona, Spain

Built between 1909 and 1911 by the architect Josep Puig i Cadafalch, the Casaramona factory was declared a national monument in 1976.

The renovation program called for its conversion into an exhibition center which, in addition to the basic exhibit spaces, would include a series of complementary rooms, such as an auditorium, media archives, halls and offices. The required surface area would be double that of the existing building.

The available space —with standardized, homogenous and versatile naves— was ideal for its conversion into exhibit halls, without having to tear anything down or undertake a major overhaul. Thus, assessment of the existing space, along with the desire to conserve it as an exhibit space and the need for more surface area, led to the scheme's central decision to house the additional functions in a new basement which would occupy the entire floor space of the factory. In order to form a coherent whole, the design for this basement was based on the existing architecture, thereby integrating the balance of the old building into the new.

Two autonomous volumes —one opaque (the reception and concierge) and another transparent (the library)— organize the space. The same idea of ordering the spaces through independent elements housing specific functions (translating booths, offices, bathrooms and stairwells) recurs in the rest of the building. Finishes have been resolved using veneers with no tectonic function, and with materials such as steel and glass, which comprise a contemporary space within the existing building.

By locating the new access in the basement, done by the architect Arata Isozaki, the main entrance has been exchanged for a more suitable one. A new areaway takes care of the necessary change in level and leads to the lobby, where the exhibit space is located.

The linearity and extensive use of white in the new entrance contrasts dramatically with the rest of the complex, thereby creating a rich, thought-provoking dialogue between the two architectural styles and bringing its style closer to that of Mies van der Rohe's pavilion, which lies just opposite. A sculptural pergola, which shelters the escalator access, presides over this space.

Photographs: Duccio Malagamba

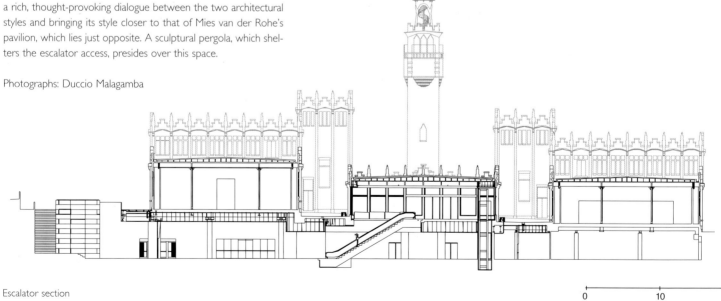

Escalator section

0   10   20

The basement translates the formal scheme of the ground floor into two large, length-wise spaces (vestibule and storage) and two central areas (auditorium and media archives). The circulation zones correspond to the inner walkways of the ground floor, to which they are connected via a central nucleus of elevators and escalators and four secondary groupings.

The organization of independent bodies enabled great flexibility of use, with the inner walkways playing a central role in the formalization of the floor plan and as support for circulation between the different areas.

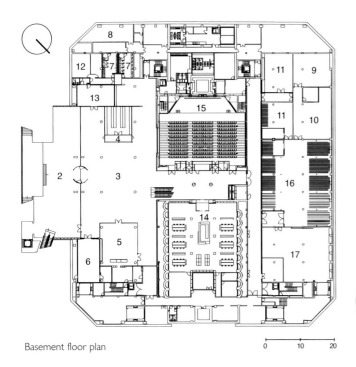

Basement floor plan

0 10 20

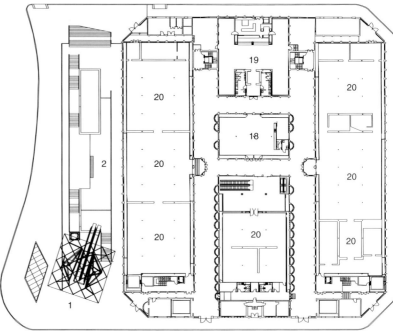

Ground floor plan

1. Access pergola
2. Open areaway
3. Hall
4. Reception and concierge
5. Shop
6. Multi-purpose hall
7. Bathroom

8. Machine rooms
9. Photography workshop
10. Restoration
11. Storage room
12. Security and control
13. VIP Room
14. Media archive

15. Auditorium
16. Storage for artwork
17. Packaging
18. Arts lab
19. Restaurant
20. Exhibit hall

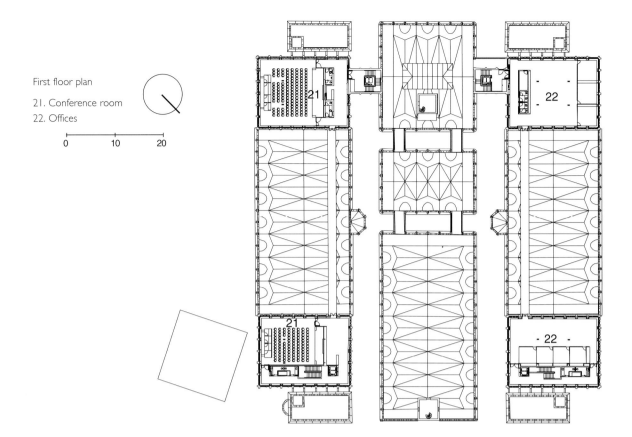

First floor plan

21. Conference room
22. Offices

0   10   20

119

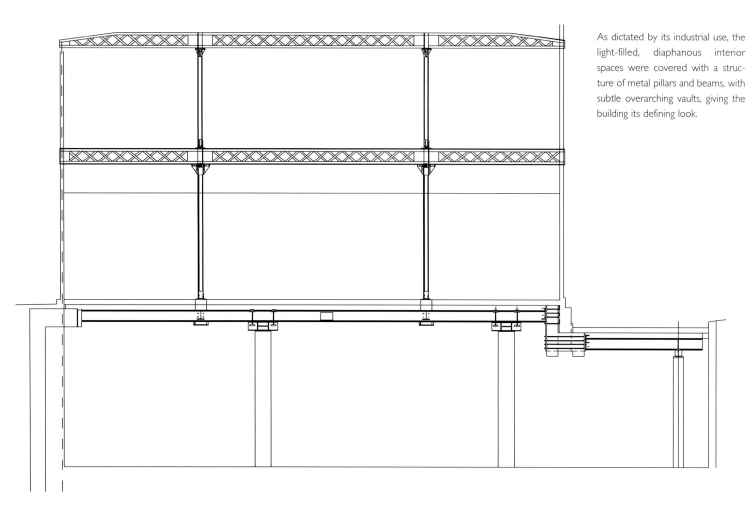

As dictated by its industrial use, the light-filled, diaphanous interior spaces were covered with a structure of metal pillars and beams, with subtle overarching vaults, giving the building its defining look.

Section of standard portico, nave B

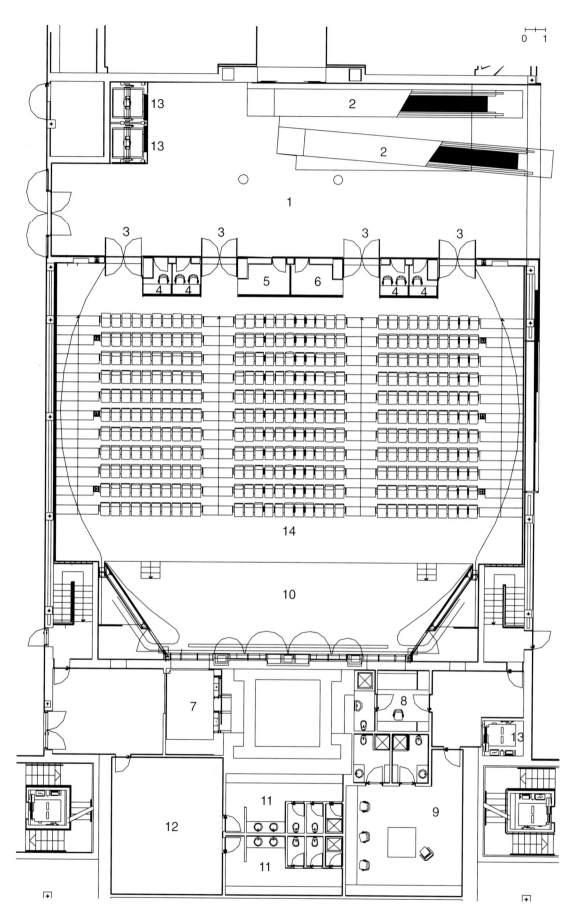

Auditorium floor plan

1. Lobby
2. Entrance escalator to exhibit room
3. Auditorium entrance
4. Translator's booth
5. Screening room
6. Sound room
7. Service elevator
8. Individual dressing room
9. General dressing room
10. Stage
11. Bathroom
12. Dressing room
13. Elevator
14. Auditorium seating

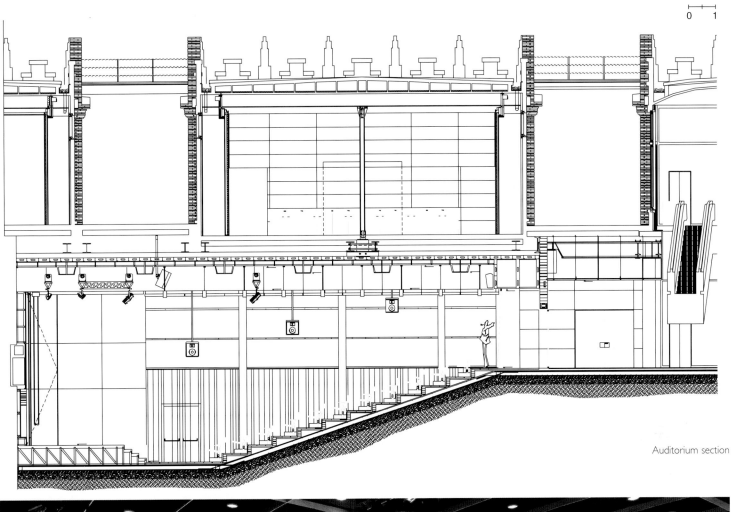

Auditorium section

Rudy Ricciotti
Montmajour Abbaye

Arles, France

The project for the creation of a visitor's center within the vaulted cellars of the 10th century Benedictine Abbaye de Montmajour was won in competition. As these Romanesque ruins are cherished for their considerable historic, aesthetic and architectural appeal, the basic idea behind the winning scheme was to respect the original monolithic structure as much as possible, creating a sort of stage set. Thus, the visitors' center acquires a transitory feel, as opposed to the timelessness of the surrounding structure.

The design team also chose to highlight the difference between the new and the old – present and past. This dichotomy is seen in details such as the blue glow emanating from fiber-optic tubes placed inside the water conduits, which are carved into the stone walls of the main hall, or in the colored and illuminated modules housing the lavatories inside a vaulted stone chamber.

The initial work consisted of cleaning the stone and recuperating openings which had long since fallen into a state of disrepair. The new underground entrance, now fitted with wide expanses of clear glass, passes beneath an elevated steel and glass walkway – steel for the necessary structure and as much glass as possible for creating unobstructed views of the building. The walkway, which is supported by a series of single columns and detached from the side walls, continues toward the exterior, where it becomes a glass encased bridge.

Of the two rectangular chambers with sloping floors comprising the visitors' center, the largest houses the entrance hall. Here, a simple, impermanent ticket booth has been devised to contrast with the solidity of the vaulted stone ceilings.

A reception desk of green polyester glass resin runs almost the entire length of one wall and, like the floating concrete slab, follows the slope of the original floor. This polished, black concrete flooring, which was poured in situ, is subtly illuminated around the edges by light fixtures placed just below the line of vision.

Photographs: Seige Demailly

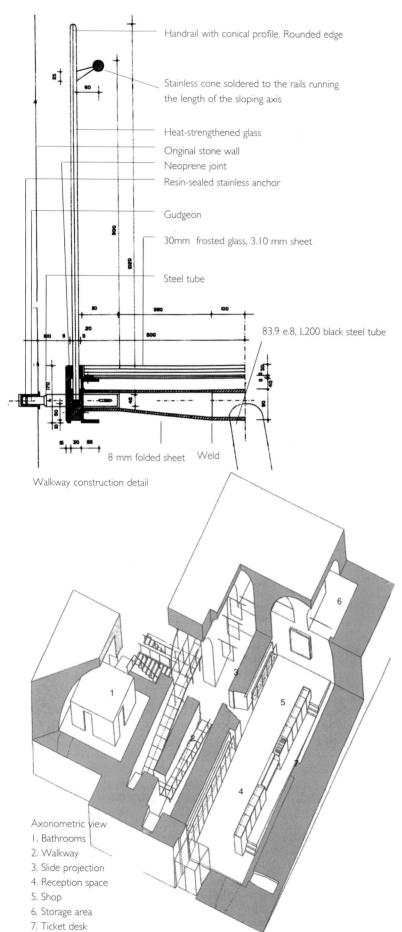

Handrail with conical profile. Rounded edge

Stainless cone soldered to the rails running the length of the sloping axis

Heat-strengthened glass
Original stone wall
Neoprene joint
Resin-sealed stainless anchor

Gudgeon

30mm frosted glass, 3.10 mm sheet

Steel tube

83.9 e.8, L200 black steel tube

8 mm folded sheet   Weld

Walkway construction detail

Axonometric view
1. Bathrooms
2. Walkway
3. Slide projection
4. Reception space
5. Shop
6. Storage area
7. Ticket desk

Fiber-optic cables placed inside the water conduits, which are carved into the ancient cellar's stone walls, are another example of the deliberate dichotomy between the old and the new.

The low, 15-meter-long desk running almost the entire length of one of the entrance hall's side walls is of green polyester glass resin.

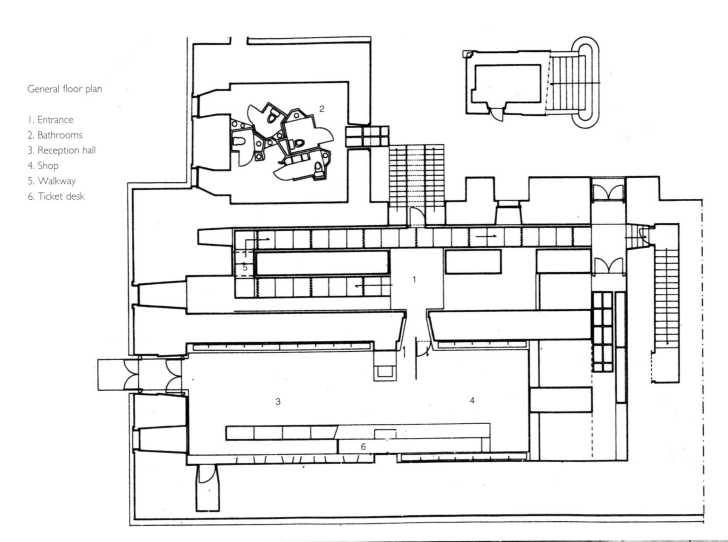

General floor plan

1. Entrance
2. Bathrooms
3. Reception hall
4. Shop
5. Walkway
6. Ticket desk

The polished black concrete floor slab was poured in situ and follows the slope of the original floor. Light fixtures are concealed just beyond the sight line along the edges of the slab, creating elegant and subtle lighting effects.

An impermanent ticket booth creates a deliberate contrast with the magnificent vaulted stone ceilings.

José Paulo Dos Santos
Inn at the Convento dos Loios

Arraiolos, Portugal

Now transformed into a luxurious hotel, the reorganization of the different functional spaces of the convent accompanies what would be its natural expansion through time. The lower floor—almost completely dug into the earth—houses the service areas with the exception of a room to be used for conferences. The main floor houses the public areas organized around the succession of exterior spaces made up by the cloisters, patio and esplanade. The upper floor houses the bedrooms, some in the old part and the remainder in the new wing.

The convent uses inexpensive materials in its built essence. Stuccoes are redone and used throughout. Stone—green xisto and granite—covers most floors. Oak is used in the flooring of bedrooms and upstairs corridor of the new wing. Local marble from unused quarries makes up the facing of bedroom bathrooms. In the interior design, done in collaboration with the architect Cristina Guedes, the oak and cherry wood furniture and other appliances have been made to measure.

Now, as before, without altering the protagonism of the existing structure—keeping all its spatial qualities intact—the addition of a new wing enclosing the eastern patio acknowledges not only the implicit formal autonomy of the existing but also the development of its own rules. These are kept in line with the character of the materials, austere but at the same time rich in iconography and forms.

Photographs: Luis Ferreira Alves

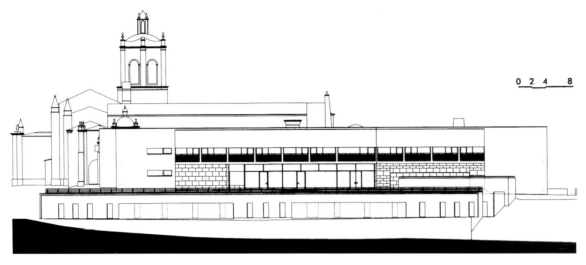

The public spaces of the convent, located on the first floor, are organized around a rhythmic succession of exterior and interior spaces.

North elevation

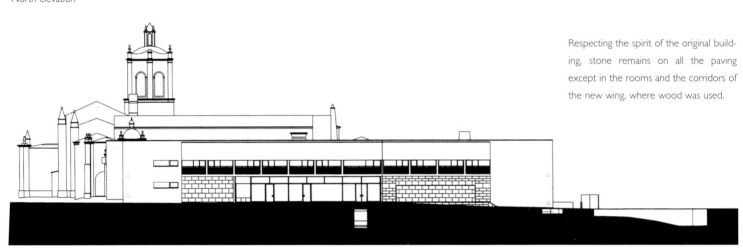

Respecting the spirit of the original building, stone remains on all the paving except in the rooms and the corridors of the new wing, where wood was used.

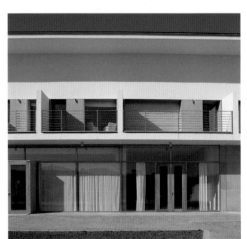

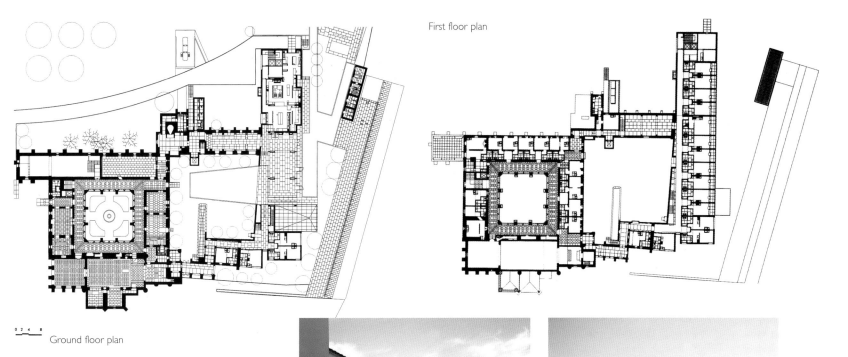

First floor plan

Ground floor plan

The photographs show several exterior views of the old convent and the extension that was added to it.

Renzo Piano Building Workshop
Conversion of the Lingotto Factory

Turin, Italy

When it was built in the 1920s, Lingotto, Fiat's birthplace and headquarters, in Turin, was the largest and most modern plant in Europe. Although it had once been the true economic and cultural symbol of urban Turin, its days as an industrial center had passed.

The premises are now used for an auditorium, an exhibition center, a branch of the university, a shopping center, a hotel and a 2600-seat cinema complex, as well as Fiat's headquarters. The concert-conference hall has been constructed beneath one of the old building's four central courtyards, the floor of which is now raised to first floor level. The floor of the former courtyard has been sunk by 14 meters, well below the level of the old building's foundations, to achieve the hall's requisite volume and sloping floor.

For acoustic insulation, the new structure of the hall is completely independent of the old structure. The concrete frame expressed on the hall's long sides supports steel beams and concrete decking which are independent of the steel beams which support the 350 mm deep concrete slab of the courtyard floor. For further sound dampening, both sets of beams are mounted on rubber pads.

Apart from the sloping floor, what has been created is a rectangular volume (lined on its long sides by two galleries, with a row of translation booths above them), into which has been inserted a balcony at one end and a fully adjustable stage at the other. The suspended ceiling of convex curved segments is also adjustable. Each segment of the ceiling, and of the lighting grids between them, can be independently lowered and raised.

The right acoustic reverberation time was also obtained by drilling holes in the walls, capturing the sound with arched galleries and shattering the echo with the use of wood, which proved to be the most suitable material for this purpose.

Photographs: Shunji Ishida, Gianni Beregno & Michel Denanncé

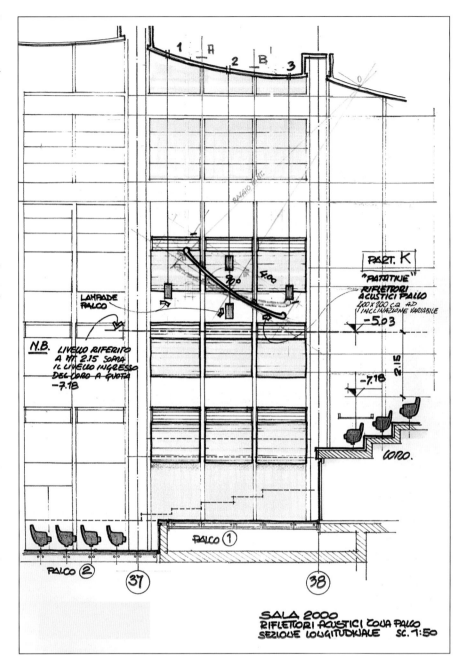

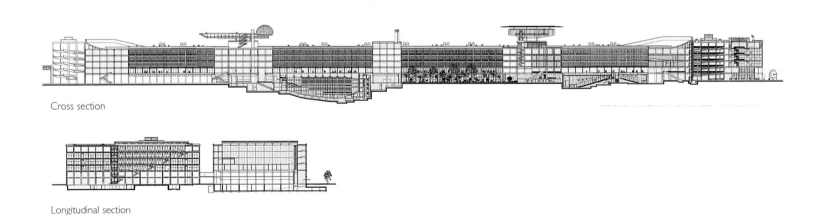

Cross section

Longitudinal section

Built in the 1920s, Lingotto was one of Europe's first examples of modular construction in reinforced concrete. The roof was (and still is) a test track for cars. One of the plant's four inner courtyards is now the site of the new auditorium. The blue bubble sitting atop the building is a conference room and heliport..

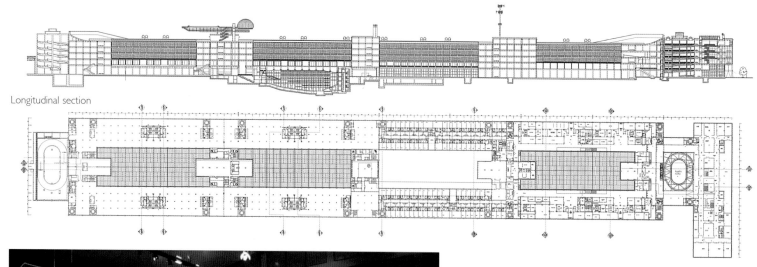

Longitudinal section

Each segment of the ceiling can be lowered and raised. At its maximum volume (24,000 cubic meters) the auditorium has a reverberation time of 1.9 seconds. The ceiling can be lowered by as much as 6m, decreasing the volume of the hall and adjusting the acoustics. Cherry tree wood, which adds a rich sound quality, was used for the flooring, ceiling panels and walls.

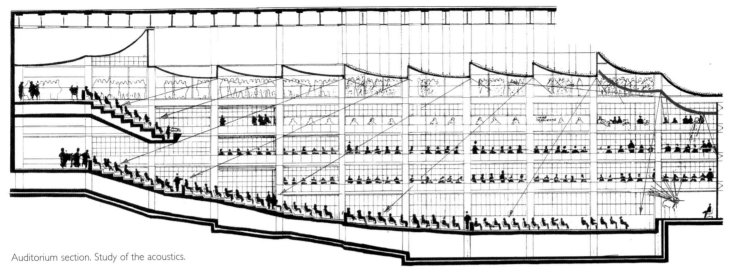

Auditorium section. Study of the acoustics.

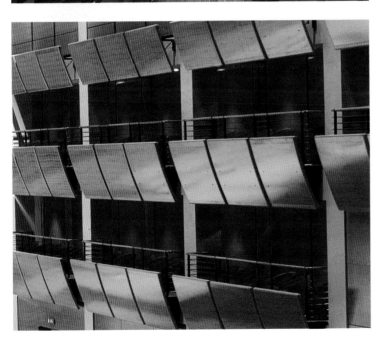

Section of balconies and translating booths (side walls).

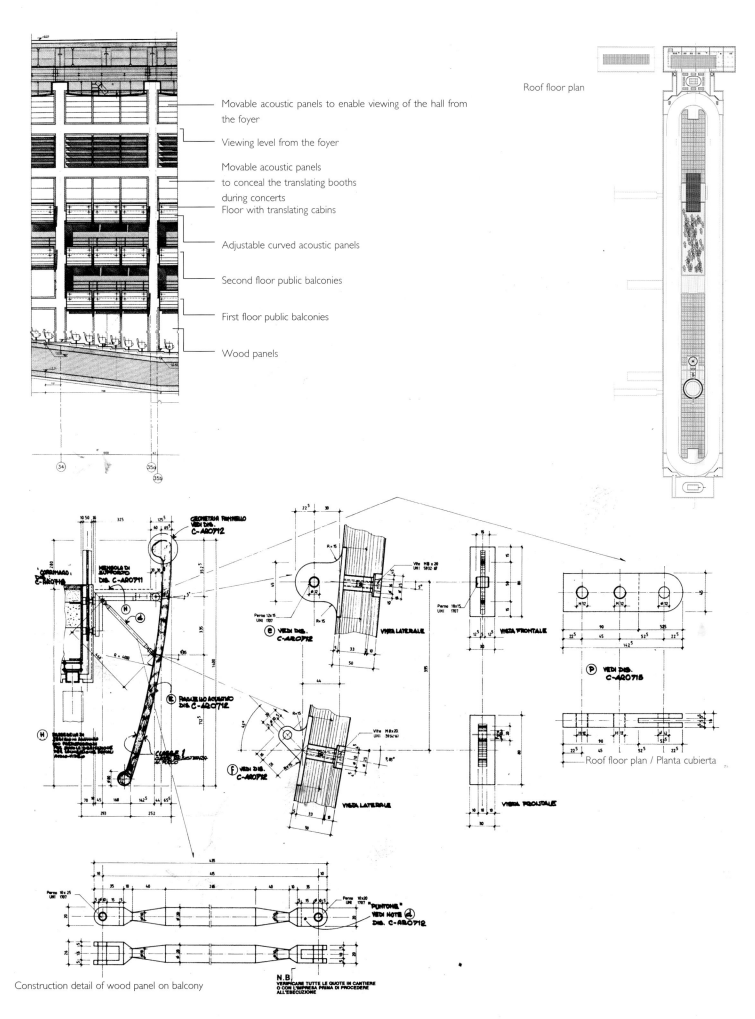

Movable acoustic panels to enable viewing of the hall from the foyer

Viewing level from the foyer

Movable acoustic panels to conceal the translating booths during concerts

Floor with translating cabins

Adjustable curved acoustic panels

Second floor public balconies

First floor public balconies

Wood panels

Roof floor plan

Roof floor plan / Planta cubierta

Construction detail of wood panel on balcony

143

Günther Domenig
Centre of Documentation
Reichsparteitagsgelaende
Nuremberg

Nuremberg, Germany

This project was an especially extraordinary one as it called for the creation of a new Documentation Center in the remains of Hitler's Congress Hall, alongside the monumental Coliseum, designed by Albert Speer.

The new exhibit space and Documentation Center is a moving "reminder-memorial" of negative contemporary history. The issue dealt with in the exhibit is intensified by the material reality of ideological architecture, which was meant to be a physical and symbolic representation of fascist strength and power, in a space intended for mass rallies and military processions.

The program essentially consists of three parts: creation of the Documentation Center and space for changing exhibits, the meeting and connection zone and the forum space for learning and teaching.

The exhibit rooms and Documentation Center are spaces for displaying the fascist architecture. The meeting zone and educational forum have been deconstructed and deprived of their original monumentality.

The existing rooms, their walls and ceilings, largely remain in their crude concrete and brick structure. The existing ceilings have been supplied with industrial floor coverings (sealed concrete screeds).

A "beam" cuts through the rectangular geometry of the northern wing, penetrating the building and jutting out over the courtyard. The entryway has been developed to include wheelchair access and an elevator has been installed.

The changing exhibit space, lecture hall and screening room are all located on the ground floor, with the Documentation Center installed on the upper floor. A cantilevered hanging terrace perches atop the building and hangs out over the top floor.

All new architectural elements have been built with steel, reinforced concrete, glossy aluminum cladding and glass. The existing walls have been left almost entirely intact, with some openings broadened in the areas requiring passage for the exhibit.

Photographs: Gerald Zugmann

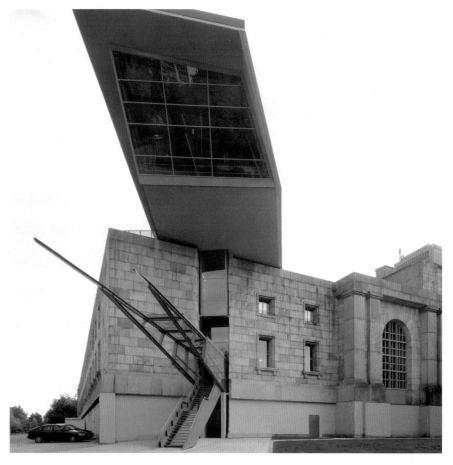

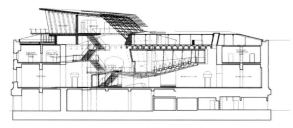

Cross section

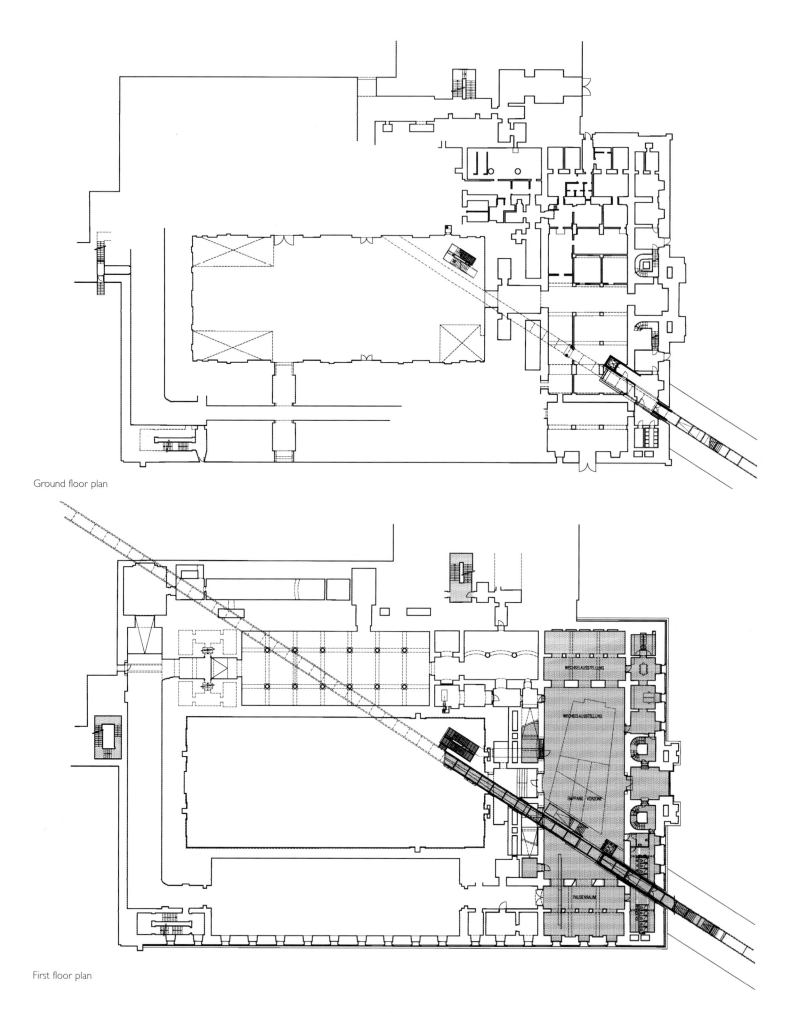

Ground floor plan

First floor plan

146

A sloping corridor has been wedged into the building, emerging on the facade to form the entrance to the new Documentation Center. It cuts through the building and ends up hanging over the courtyard on the other side. The original structure was meant to symbolize fascist power, while the renovation boldly refutes it.

The unadorned concrete and brick of the original unfinished structure remains untouched; while the materials used in the renovation — steel, aluminum and glass— serve as a visual contrast between old and new, past and present.

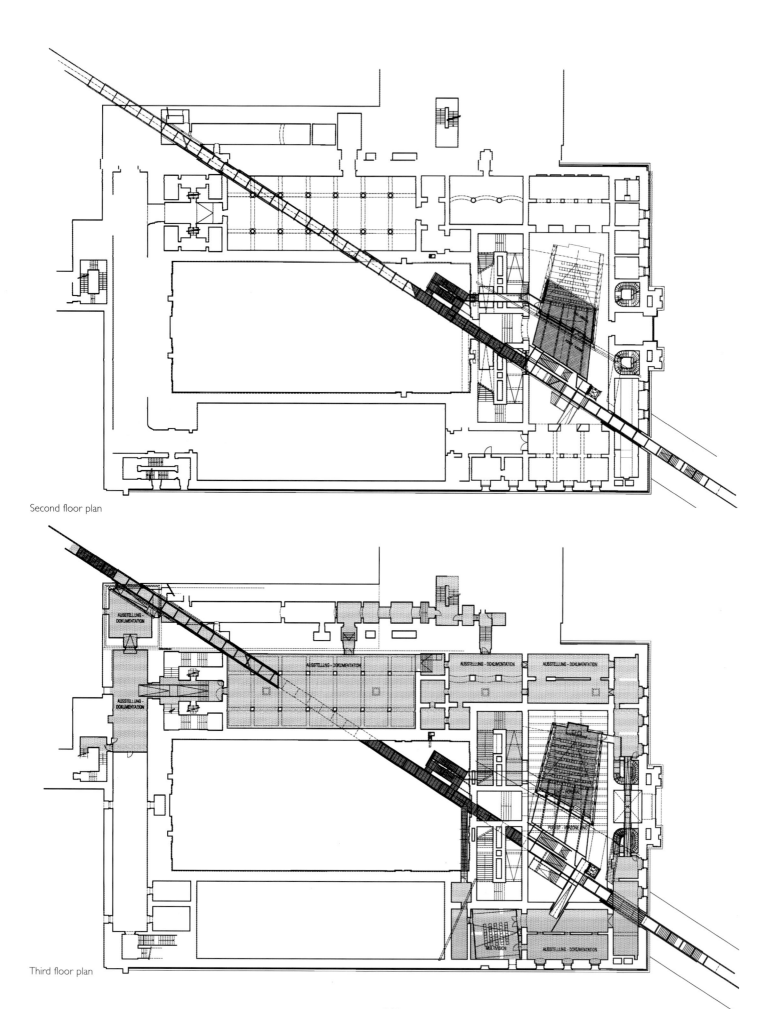

Second floor plan

Third floor plan

148

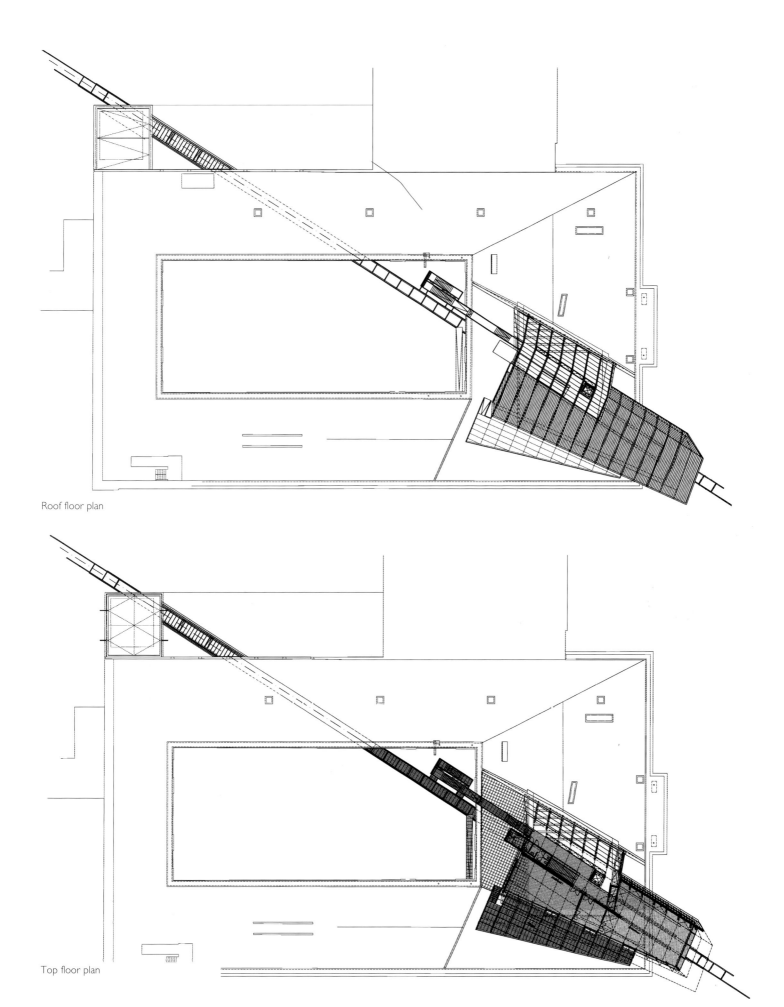

Roof floor plan

Top floor plan

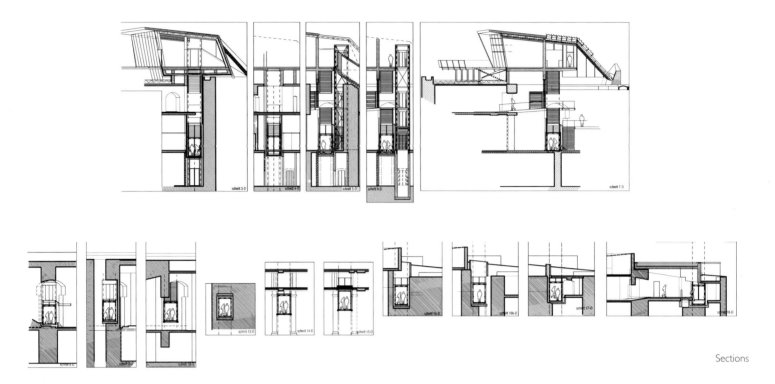

Sections

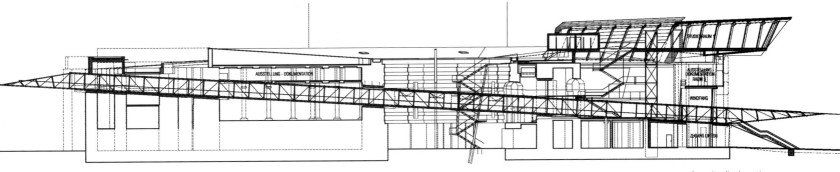

Longitudinal section

151

De una forma u otra, todos los espacios se han "deconstruido" para eliminar la sensación de monumentalidad deliberadamente impuesta por la construcción original. A pesar de ello, la renovación ha mantenido intacta la arquitectura original con fines históricos y educativos.

Klaus Sill & Jochen Keim
Apartment/Office Building

Rathenow, Germany

This building was built over a century ago, although it was later modified by joining the front area to the courtyard. Now, the building has been integrated homogenously in an environment that is distinguished by the walls of the block and the intelligent conception that makes it suitable for working and living.

The courtyard building had fallen into a state of ruin in the last few decades: some walls had collapsed and others were in danger of collapsing.

It was decided to only conserve an old dwelling and a warehouse, two elements located at the north-west end of the land that were considered to be worth maintaining.

As part of the project, a major installation was built inside the block to serve as a landscaped area and playground for the tenants of the front building. Also, thanks to this arrangement, it was possible to cover the objectives agreed in the urban plan of the city of Rathenow concerning the separation between the blocks and the foundations.

The conjunction between offices and dwellings was achieved by means of a distribution in which the upper floors were reserved for the dwellings. Thus, the ground floor and first floor contain the cubes (12 modules) that house the offices of a firm of engineers with twenty employees. The second floor is composed of three maisonettes of about 60 and 90 m².

For the firm of engineers two united surfaces were used because the building protrudes about 4.5 m into the courtyard. This extension is formed by two conferences rooms, an audio-visual room, an archive, a kitchen and a toilet.

Photographs: Christof Gebler& Klaus Sill

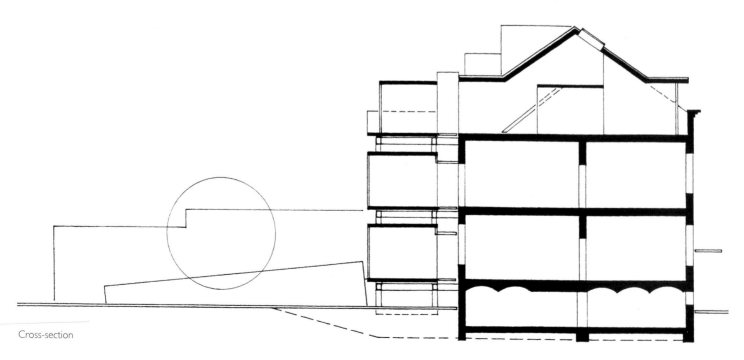

Cross-section

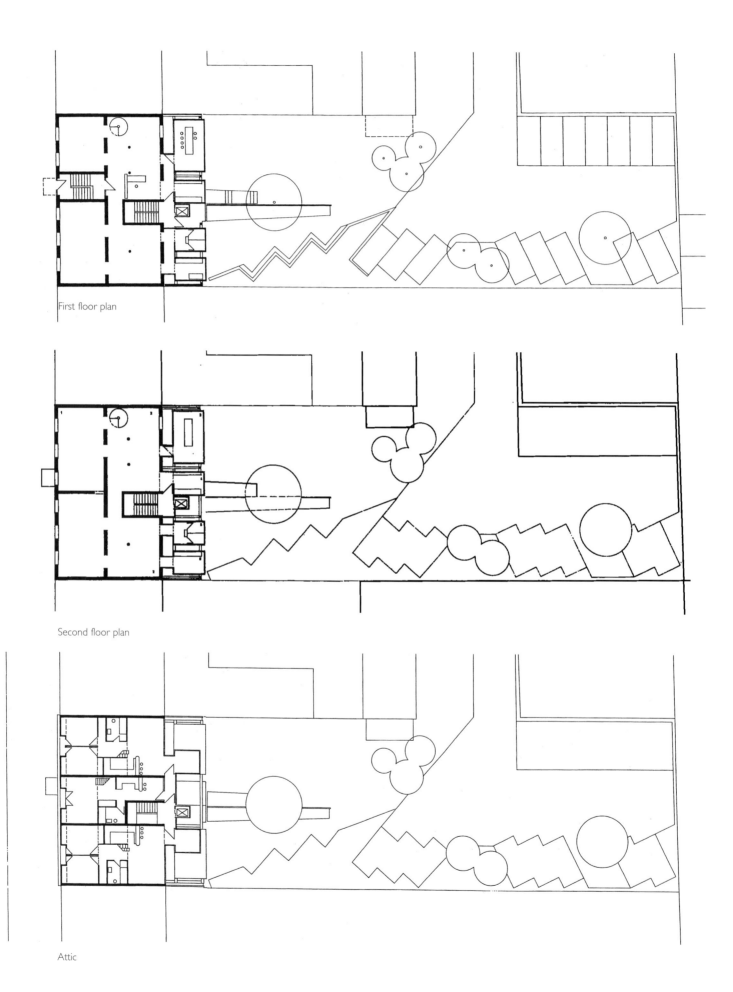

First floor plan

Second floor plan

Attic

158

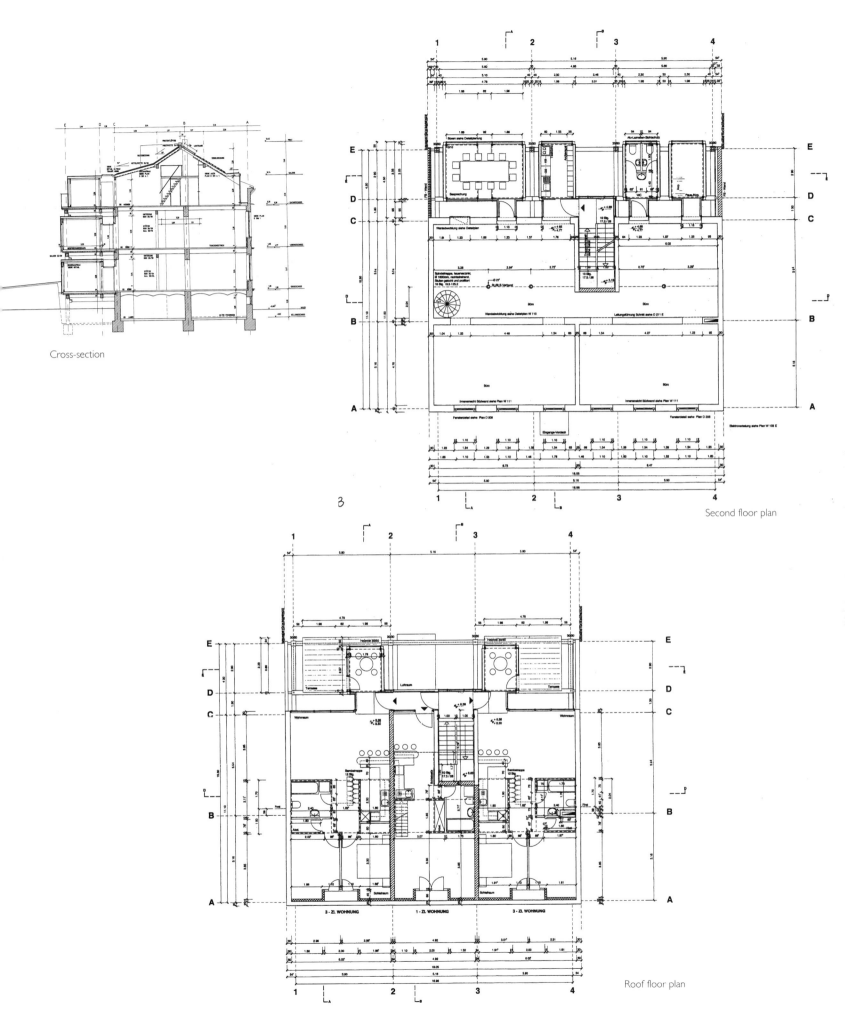

Cross-section

Second floor plan

Roof floor plan

159

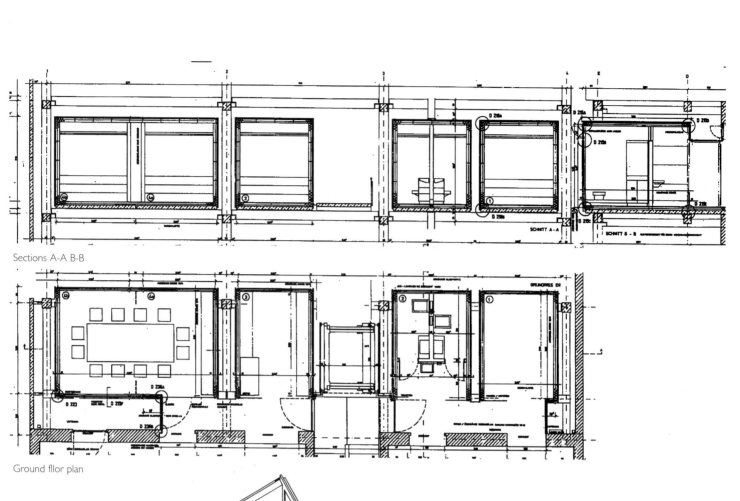

Sections A-A B-B

Ground fllor plan

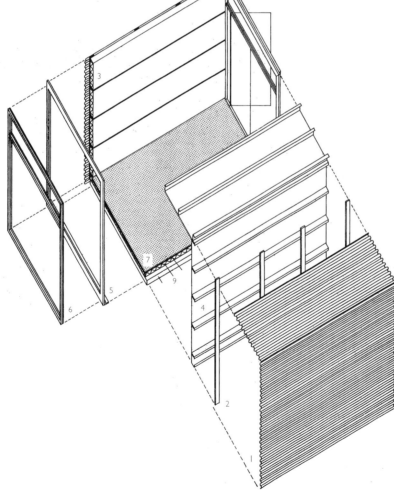

Axonometric view

Construction of the roof and walls

1. 20 mm corrugated aluminum
3. 120 mm thermal insulation
4. Waterproof layer
5. 120 mm steel plate floor
6. Facade of stiles and rails
7. Insulated glass

Construction of the floor

8. Linoleum
10. 80 mm thermal insulation

For economic and technical reasons, these modules were built 500 km from the site because they were of a suitable size to be transported.

They were set into a prefabricated reinforced concrete skeleton for fire protection.

The project takes all its characteristics from the building and from the environment in which it is located, and it integrates the mixture of dwelling and workplace in a way that is more in line with the current lifestyle.

Toni Cordero
Factory in Via Teatro Vecchio

Mantua, Italy

The complex in which this project is developed, in Via del Teatro Vecchio, Mantua, is one of the oldest constructions in the Italian city and has a privileged location next to the Teatro della Comedia. The old plans show that this was an area of great importance in the middle ages due to its proximity to the seats of political and religious power. The area also enjoyed great prestige in the early years of the 17th century at the time of the court of the Gonzagas.

The architect Toni Cordero, in collaboration with the architects Corrado Anselmi and Antonia Pintus, undertook the task of transforming these delicate remains into a new dwelling.

Due to the many transformations and additions that these constructions have under-gone over the centuries, the physical and functional quality of the complex had deteriorated. With the exception of the two modern bodies of the complex – which have no artistic value and were therefore used for the service areas of the dwelling – the intervention concentrates on the remaining old elements, and is governed by a total respect for the original layout. In order to avoid altering the walls and the originality of the old ground plan, it was decided to give each room an outer wall of colored cement, tracing a new layout without damaging the old division. This system would also facilitate a future restructuring of the space without the need for complicated work. In the case of the floors it was not possible to respect some parts because all the original materials had disappeared. They were therefore totally redone using two types of paving in the whole of the house: colored cement with iron strip joints, in some cases with inserted pebbles that illustrate motifs of the history of the Gonzagas, and floorboards with a certain industrial flavour.

Photographs: Santi Caleca

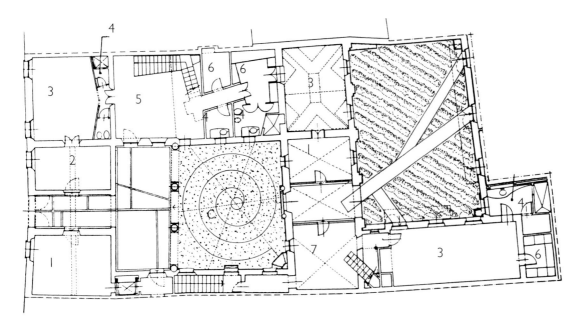

Ground floor plan

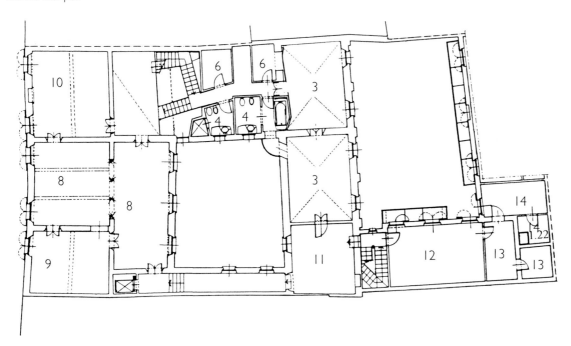

First floor plan

1. Study
2. Restroom
3. Bedroom
4. Bathroom
5. Main entrance
6. Storage room
7. Service entrance
8. Living-room
9. Dining-room
10. Library
11. Office
12. Kitchen
13. Storeroom
14. Laundry

As part of the rebuilding process, the pavement has been restored by means of an application of colored cement and interspersed iron strips that serve a dual purpose: they act at the same time as an expansion joint and as design elements which evoke the world of the Gonzagas in the city of Mantua.

The interior distribution is subtly displayed as a sequence of spaces arranged in order of importance. This system achieves various levels of privacy, ranging from domestic intimacy to the vitality and communication of the more public spaces.

168

Thanks to an intelligent distribution of light and a delicate cladding of industrial wood treated with lacquered parquet on the floors, the rooms of the upper floor acquire a certain air of theatricality without losing any spatial warmth or comfort.

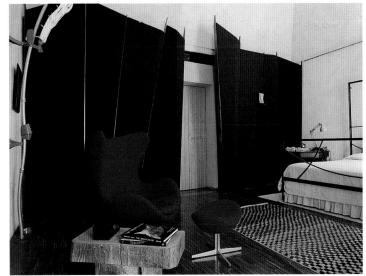

Guillermo Vázquez Consuegra
Andalusian Institute of Historical Heritage

Seville, Spain

La Cartuja de Santa María de las Cuevas can be regarded as a miniature city standing opposite Seville and also fenced in by it, on the banks of the River Guadalquivir. Like any urban entity, it has undergone a continuous process of modification over its five centuries of existence. A dense network of kilns, chimneys, bell towers and spires rise over the island as a witness to its checkered history. Originally, in the 15th century, it was a Carthusian monastery; only becoming a famous pottery in the first half of the 19th century under the management of the English entrepreneur Pickman. The complex became increasingly chaotic and labyrinthine, the more recent industrial structures blending in with and superimposed onto the earlier religious buildings to create a unique pattern of relationships between the two.

This was the situation when part of the complex was given a thorough restoration for the Universal Exposition of 1992, which was held on the island. Three year later, another section was inaugurated to house the offices and workshops of the Heritage Institute of the Department of Culture of the Andalusia government.

The project for this last stage, by the architect Guillermo Vazquez Consuegra, focuses on the so-called Manufacturing Area, characterized by industrial installations and featuring few religious elements. The scheme forms the basis of considering the sector as a conglomeration of parts. It emphasizes its episodic, discontinuous nature and attempts to construct its edges appropriately, adding new building and completing the fragments, while at the same time respecting the unique urban quality of the original building with its cloisters, alleys and catwalks.

The first stage of the construction work was carried out without a program of usage; so it was the remains of the old buildings which would suggest the eventual direction to be taken. In this way, a new architecture was proposed that sought its origins in the experience of that which already existed. However, the architect stressed the importance of not overestimating the remains or the mere fact of their antiquity but of their architectural, constructional and historical value. Some elements were therefore demolished, only those considered to be of quality being preserved. The intention was to create an architecture free of formal and stylistic mimicry, capable of inserting itself naturally into the long process of growth and transformation of this group of monumental buildings.

Photographs: Duccio Malagamba

First floor plan

Ground floor plan

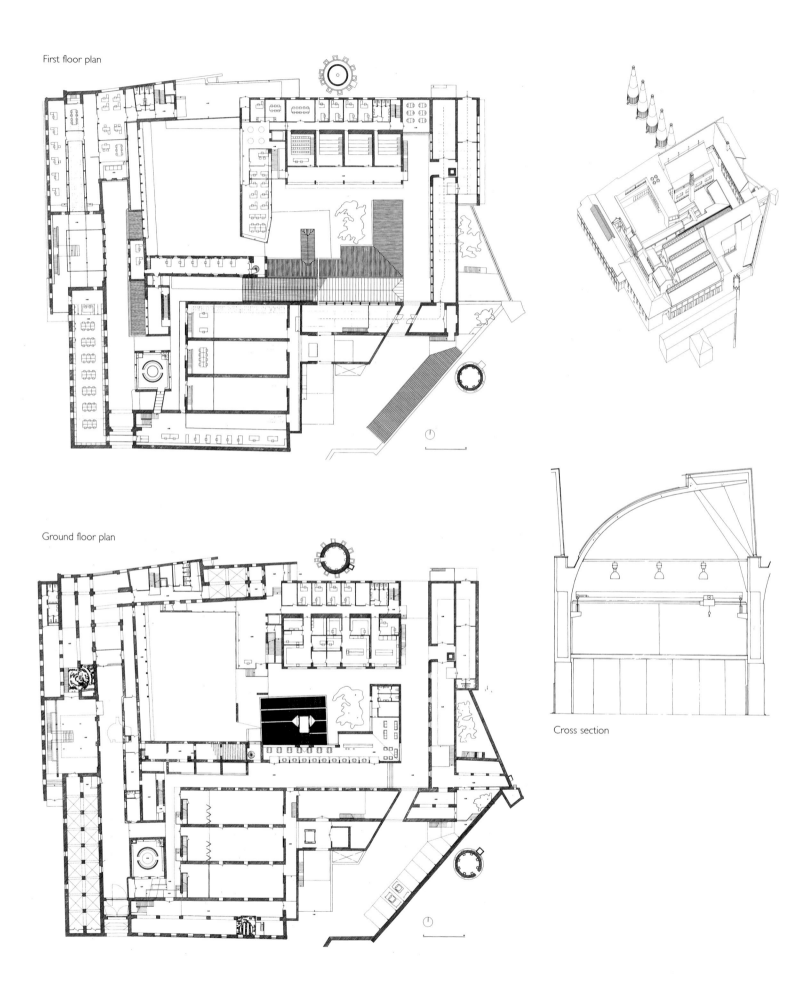

Cross section

172

173

Section
through the library

Section though
administration area

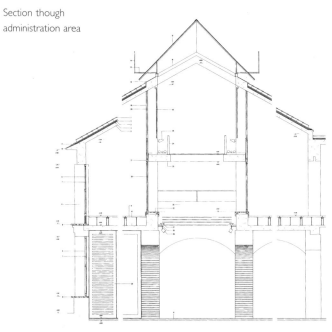

Longitudinal sections

The interior of the complex is reached through a large archway. Only high-quality existing architectural features were preserved, and those lacking any value were removed.

Francesco Delogu & Gaetano Lixie
Catrani Catrani

Umbria, Italy

Restoration of the *Castello di Petriolo* in a valley not far from *Cittá di Castello*, in the Italian district of Umbria, has resulted in refurbished interiors, which are fully adapted to modern living requirements and display total respect for the historical fabric of the building itself.

Built in medieval times as part of he nearby town's defensive network, the castle complex has numerous architectural stratifications testifying to the variety of uses it has been put to over the centuries, from noble residence to farm house.

In 1736, Marco Antonio Catrani, counselor of th Roman Curia, redesigned the main façade, making two large bulwarks to access the courtyard, and some interior modifications.

The recent project by Delogu and Lixi focuses the conversion of the complex into a set of private dwellings. So they made a general conservative restoration plan and organized its division into four separate apartments.

The so-called chapel apartment featured here occupies only part of the wing to the left of the main portal. Its three-level design incorporates the previous layout without overwhelming it, creating beautifully contrived contrasts between austerity and complexity.

Photographs: Roberto Bossaglia

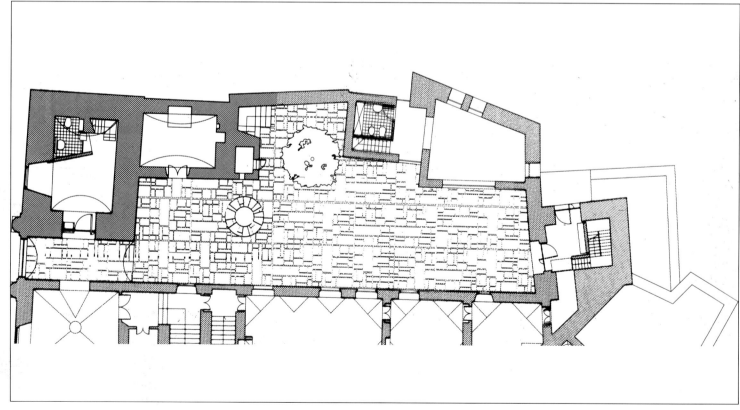

General floor plan

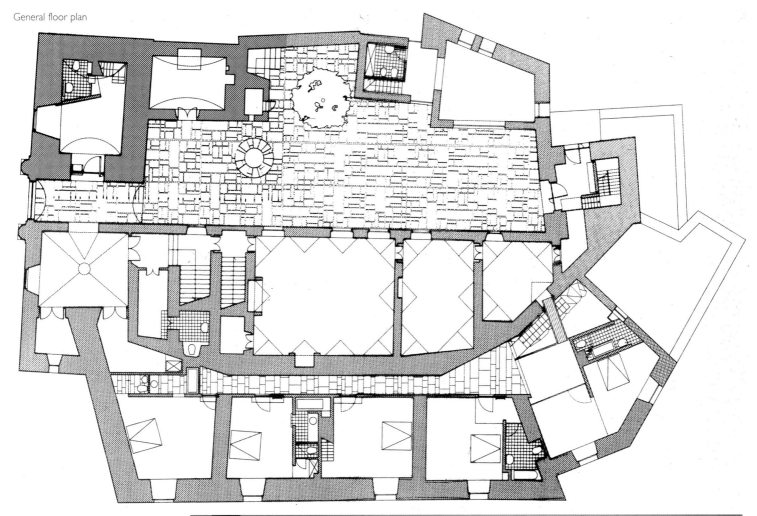

The entrance courtyard of the castle, the restoration of which was limited to conserving existing elements. The courtyard is flanked by the "chapel apartment", so called because the interior incorporates the former chapel of the castle.

Ground floor plan

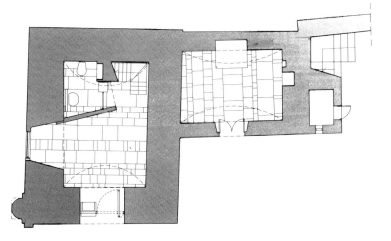

First floor plan

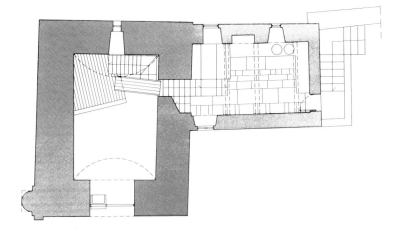

Second floor plan

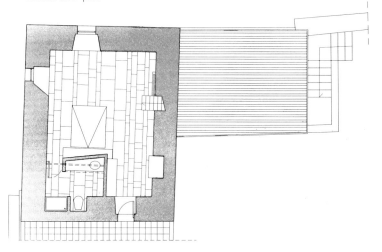

Louis Kloster
Sola Ruin Church

Jaeren, Norway

The project is focused on the reconstruction of a small Romanesque church, built in open countryside near the sea. The architect wished to capture the spirit of the location, the essence of the small construction in the immensity of the surroundings.

The Romanesque church was rebuilt stone by stone. By juxtaposing the spirit of the Middle Ages and the contemporary period, it is easier to discern the contrasts in building customs and technologies, and the differing interpretations of light and dark.

The original natural dark diabase stone is used for all the walls. The joint heights in the overlaps, arches and openings have all been constructed in part based on markings and measurements from the last pre-demolition survey. To retain some of the character of the ruins the missing stone blocks are sometimes replaced by glass tiles. This also provides a deliberate articulation to the way the light is falling.

The roof construction has been recreated in the spirit of the building with massive oak timbers and interior boarding. The new floor is made in slate and conceals heating cables.

The altar is crafted from a large, rectangular stone block excavated from the foot of the tower. Lighting for the winter nights is from the simplest possible small, cylindrical pendulums.

This church used to be a closed space providing protection from the weather and natural forces, a space for contemplation and prayer. Today it is a richer space offering contact with the elements and our wider understanding of the universe.

Photographs: Lui Costa

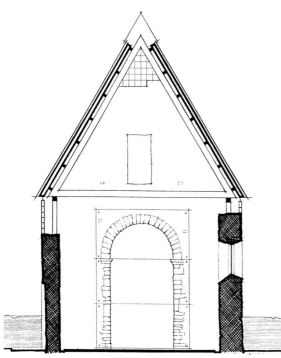

Cross-section

Site plan

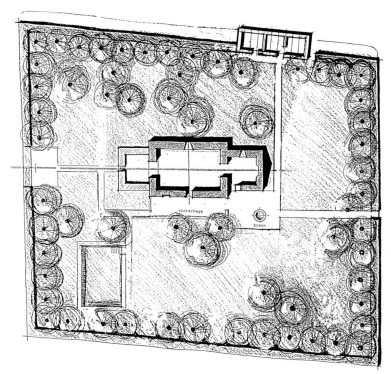

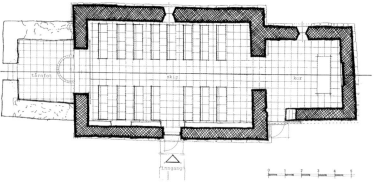

The Romanesque church stands on a platform in the middle of a landscape dominated by the fields and the sea.

During the restoration process the original stones were used wherever possible. Where they were not found, it was decided to use glass blocks, as at the top of the west facade.

The new structure covering the building was made wih large oak boards with a continuous strip of glass forming the ridge.

In the places where it was not possible to find the original stones, they were replaced by glass, allowing a great amout of light to enter the interior space.

Sudau, Storch & Ehlers
Alte Nikolaischule

Leipzig, Germany

In 1990, the Alte Nikolaischule was standing derelict and unused. Nothing remained of the glory of Germany's oldest bourgeois school, which had been founded in 1512.

Some years ago its rehabilitation and future status as cultural attraction was decided upon, with a program of demanding proportions. On the ground floor, a café was to serve as a cultural meeting point; a collection of university equipment was to be displayed in the cellar, while an antique art collection was to be located on the first floor. Above this, a story was to be provided with lecture and study rooms, in which the principal role was to be assumed by the great hall, whose basic features had been preserved. The uppermost story was destined for the Saxony Academy of Sciences.

The architects decided to embrace old and new, to rebuild between the contrasting aspects and to generate visual tension, which would fuse the contrasts together and create an identity of its own.

The renovation concept was aimed at reconstructing the external appearance and the historical interior. One after another, historically significant features emerged. The large room on the ground floor of the central house was revealed to be the school auditorium. Above the entrance hall, a beautifully painted wooden ceiling dating from the Renaissance was discovered. And on the upper floors colored plaster panels appeared. However, too many changes had been made to the building, which made a faithful restoration difficult.

In their design for an access zone to the rear of the building, the architects applied a completely new architectural syntax, creating a light, glass-roofed atrium. The upper level exhibition rooms and the attic offices are reached via two stairways and an elevator set in an imaginatively designed space.

Photographs: Angela Otto, Friederich Ostermann

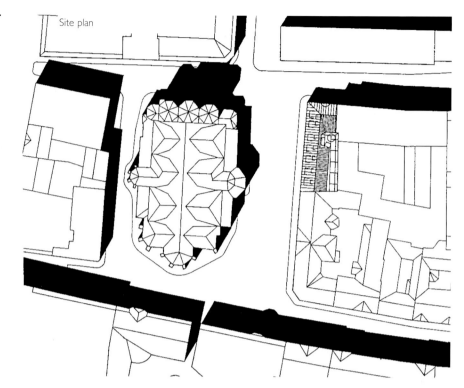

Site plan

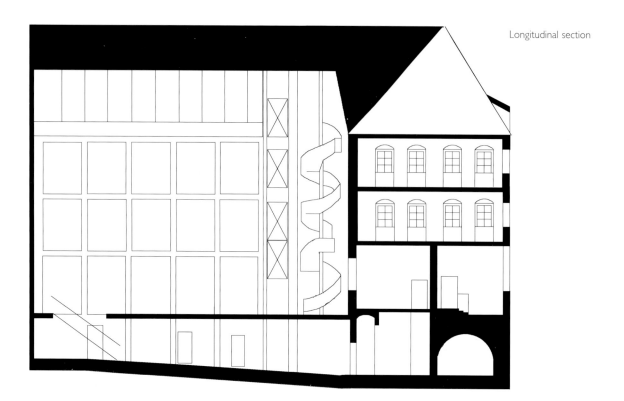

The access zones are a good example of how the design of the school creates a close relationship between old and new.

The two stairways located in the atrium and illuminated by the large glass opening act as a link between the different parts of the building.

In designing a new access zone, located at the rear of the building, the architects used a thoroughly new architectural language.s

The rear of the building is organized spatially around a large atrium illuminated from the top by a glass roof.

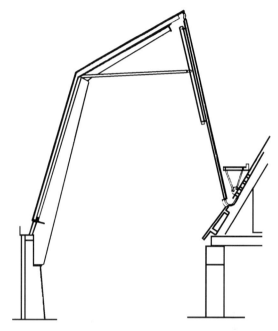

Construction detail of the roof

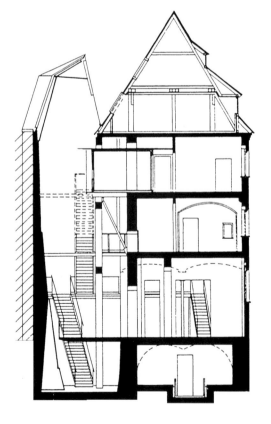

Cross section

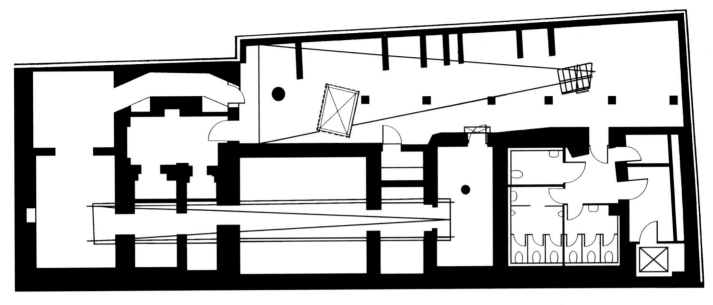

Basement floor plan

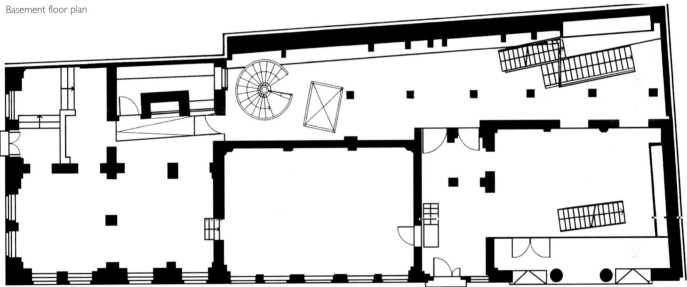

Ground floor plan

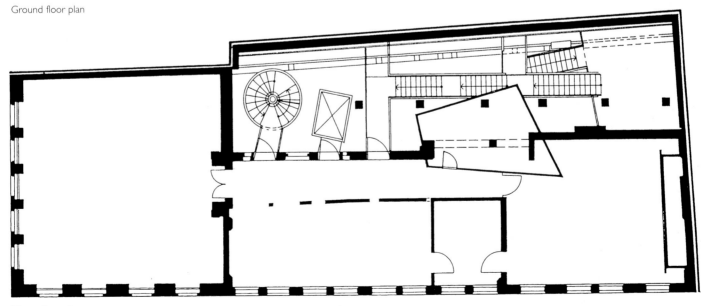

Upper floor plan

Adolf Krischanitz
Kunsthalle Krems

Krems, Austria

The new art gallery occupies a side and the interior of an abandoned tobacco factory. On the ground floor, roughly rendered and steeply sloping columns beneath short-span vaults constrained the space. Above this, two rows of columns, some of wood, some of cast iron, create three aisles through two production shops.

Krischanitz´s scheme treated the structure of the existing building with great care. The new concrete stele in front of the entrance is the only visible indication that anything has changed inside the building. The yellow of the walls and the brown of the windows relate to the world of tobacco. The careful juxtaposition of old and new is also seen in the fact that the drainpipes of the new parts of the building are under the eaves of the old building.

Krischanitz places a large up-ended cuboid in the courtyard, a new stone element which creates a space through its relationship to the old building. It has a mezzanine with an exhibition hall that has perfect environmental control and is lit from the sides by high windows which can be blacked out. Under this is a stepped lecture theater. A two-story service corridor on one side and a set of ramps on the other provide connections to the old building. Together they surround the courtyard which is now smaller than before and has a glazed roof forming a top-lit atrium.

The color scheme of the new parts of the building is based on the gray of the exposed concrete. The spatial density of the entrance hall contrasts strikingly with the spaciousness of the large atrium with its glass roof.

From the ramps visitors can look through a row of slender columns into the hall, and then as they climb higher they can look down onto it.

Photographs: Gerald Zugmann

Section

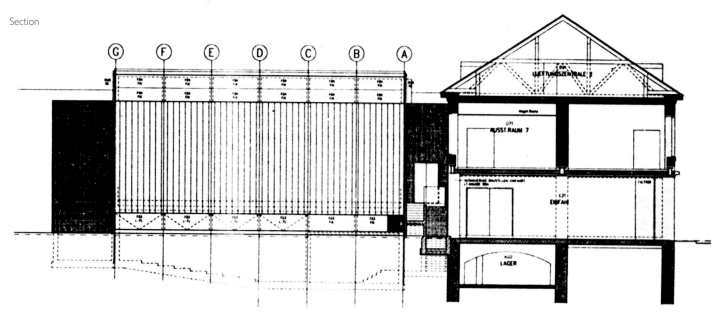

The old building is connected to the new extension through a set of ramps and a service corridor located at both sides of a top-lit atrium.

The architect has extended the useful surface of the building through the addition of a large cubic volume of glass and concrete that energises the dialogue with the old construction.

200

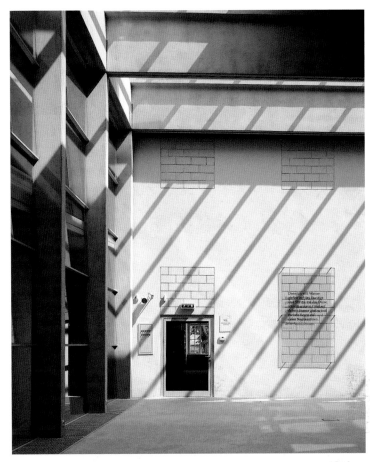

The new art gallery is located in an old tobacco factory. The architect has extended the useful surface of the building through the addition of a large cube-like volume of glass and concrete that energizes the dialogue with the old construction.

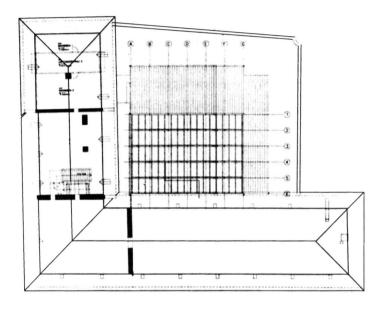

Basement floor plan

Ground floor plan

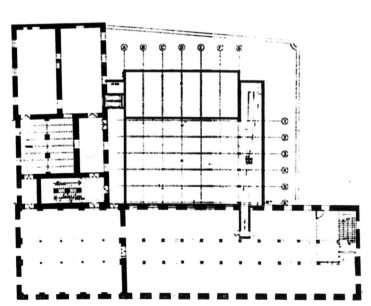

First floor plan

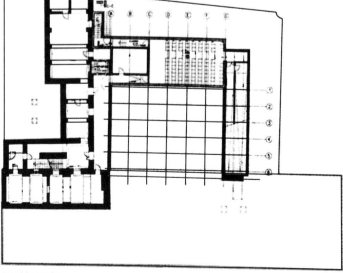

Roof floor plan

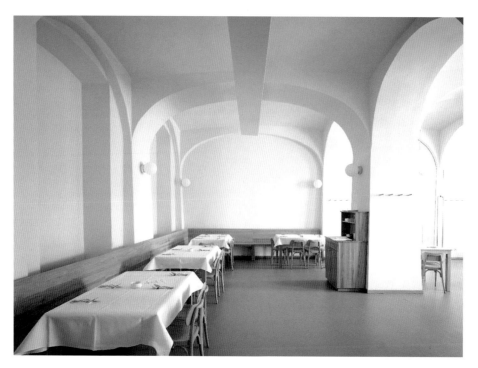

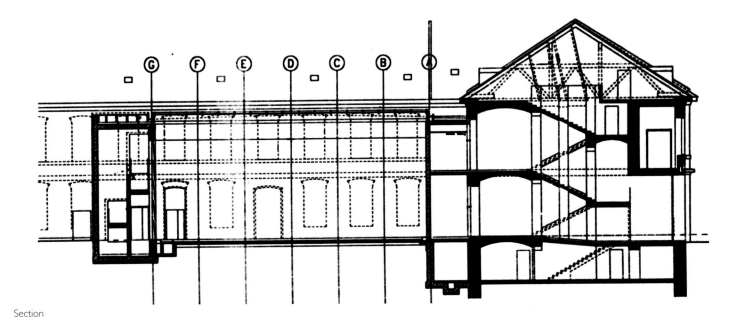

Section

Adrien Fainsilber & Associates
Nouvel Hôtel de Ville de La Fléche

Sarthe, France

The multiple program was to design an administrative building and council chamber building, to renovate the Château des Carmes and the entrance pavilion, and to design the public spaces surrounding the City Hall. Including a city square, a water cloister, a lower square and a footbridge to the adjoining public park.

The site of the City Hall, along the bank of the river Loire, is virtually an island, the entrance being a restored cloister built above the water. The newly created urban spaces are interconnected and linked to the city as well as to the existing squares and park by covered passages, steps and a footbridge.

Fainsilber, in collaboration with the architects Roland Korenbaum and Philippe Bodinier, has managed to integrate simple forms into the landscape by means of contemporary construction techniques that use glass to maintain the strong presence of the environment and to establish the relationship with the existing environment.

The main building is constructed partly over water, partly over land, in order to offer maximum transparency and views over the park and the Loire. The characteristics of the island determind the layout of the design. The curve of the east facade hugs the contour of the banks.

The reconstruction of the cloister gives coherence to the old and new buildings, and creates a new main entrance for the City Hall. The desire was to open it toward the public and to make it the point of union between the town center, the river and the park. The Parvis, the cloister and the lower square are three new public spaces on the promenade offered to pedestrians between the town center and the new views of the canal. These spaces are strongly differentiated and in close relation with the surrounding nature. Two foot bridges over the water and a staircase between the cloister and the lower square create articulations between these new public spaces.

Photographs: Stefan Couturier

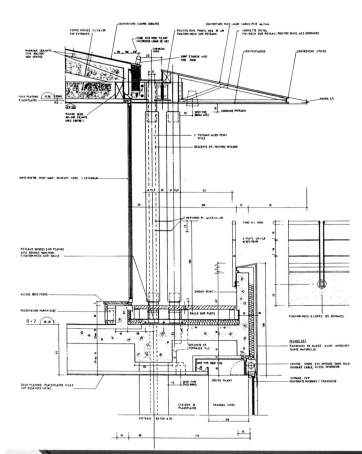

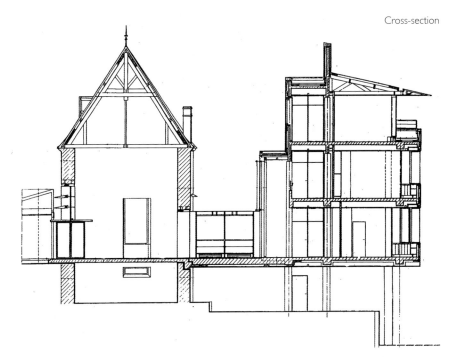

The council chamber building is partially constructed over the Loire. This situation gives the building privileged views of the river and the surrounding countryside.
The council chamber is joined to the administrative building by a fully glazed overhead walkway.

Erick van Egeraat
ING Bank & NNH Head Offices

Budapest, Hungary

The client, Nationale Nederlanden, chose a 19th century building for its Budapest headquarters. For its rehabilitation, the company chose the Dutch architect Erik van Egeraat. An approach was chosen to deliberately combine a reconstructive restoration with new additions.

The truly authentic parts of the richly ornamented Italianate building were painstakingly restored. Joining these, the new cellars, the glazed roof and the "whale-shaped" boardroom constituted the foreign elements whose contemporary architectonic elaboration contrasts starkly with the existing building. The design does not attempt to antagonize the old or the new, but combines them via the clear structure of the building and the fresh new light coming from above.

The Whale itself contains a boardroom and a coffee-corner. By the use of irregular curves, it was possible to freely form the Whale with regards to a diversity of internal spatial wishes. Its appearance therefore becomes powerful but not overbearing. External factors, such as the spatial counterform and the desire for southerly light to penetrate deep into the staircase, inspired an image of a modern lantern. It is built of 26 unique laminated timber frames that are hung on the main steel load-bearing structure, which in turn supports two concrete floors. The skin is analogous to shipbuilding, built up to orthogonal battens. On the outside it is finished with zinc and on the inside with linen. The transparent part of the Whale is made of naturally colored curved glass. The laminated glass beams of the roof support the transparent sea of clear glass in which the Whale comfortably floats.

The project as a whole is thus a manifesto against carping conservatism. It is also a message in favor of reinterpretation using contemporary means without automatically resorting to high-tech.

Photographs: Christian Richters

The roof level is dominated by a totally glazed organic mass (the "whale") that obviously challenges the
symmetry of the existing building.
The intervention was born of the firm desire to destroy any similarity between the old and the new. The
project seeks to be everything that the old building is not.

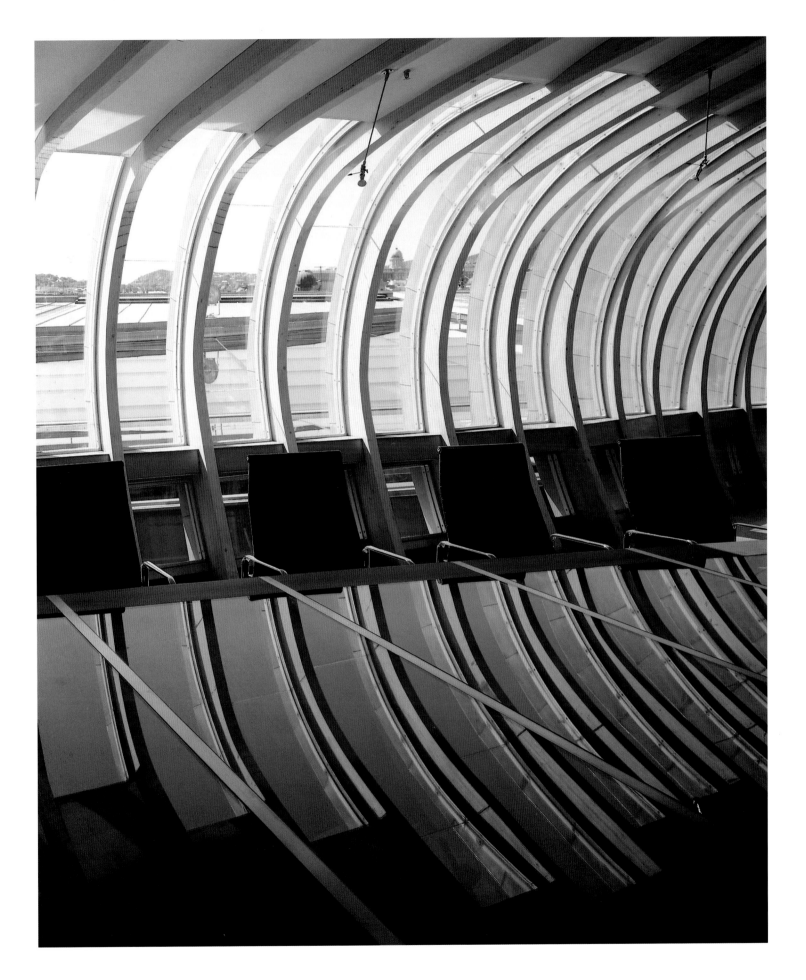

As on the outside, the inside of the building is dominated by the "whale", an element that penetrates violently into the upper level and seems to defy gravity. Functionally, besides enlarging the space, this enormous element gives light to the top floors of the building.

Gerhard P. Wirth
Loft Nürnberg

Nuremberg, Germany

The headquarters of the architect Gerhard Wirth's studio lie within an industrial complex in Nuremberg, in an old center for producing sheets of metal and zinc which later became a zipper factory. After its 1999 restoration, the old factory was made into a diverse, many-sided space which houses the 480 m^2 work area, a bedroom unit and the architect's loft which occupies 120 m^2. A new extension was added to the east facade in order to adapt the building to the needs of the new owners. This enlargement was done with a structure of wooden crossbeams, between which are layers of 20 cm mineral wool insulation. Sheets of steel were used for the exterior cladding of the new volume. OSB boards were placed between this cladding and the insulation.

The end result is a two-floor construction. The lower level houses a recreation room, the scale model room, the kitchen, the toilet and bathroom, the boiler room, a multi-purpose room and the loft. A megalith rises from this level to the upper floor.

The first floor, where the entrance to the building is located, harbors a space for group work, a terrace, a meeting room and a sizeable area designed for projects. The latter has a diaphanous feel, with a high ceiling and an abundance of natural light which streams in from the large windows and the skylights fitted into the ceiling. Along with the large windows, these skylights not only constitute an important source of natural light, but they also help reduce energy consumption. The plumbing and power installations have been left bare, thereby sparking remembrances of this building's industrial past. The will to create a flexible, versatile space was a prime consideration in the design concept, resulting in a program that would allow for the interior distribution to be adapted for new uses. Another criterion which influenced the project was the search for ways to achieve a building with sustainable energy.

Photographs: Gerhard P. Wirth; Karin Heßmann / Artur

225

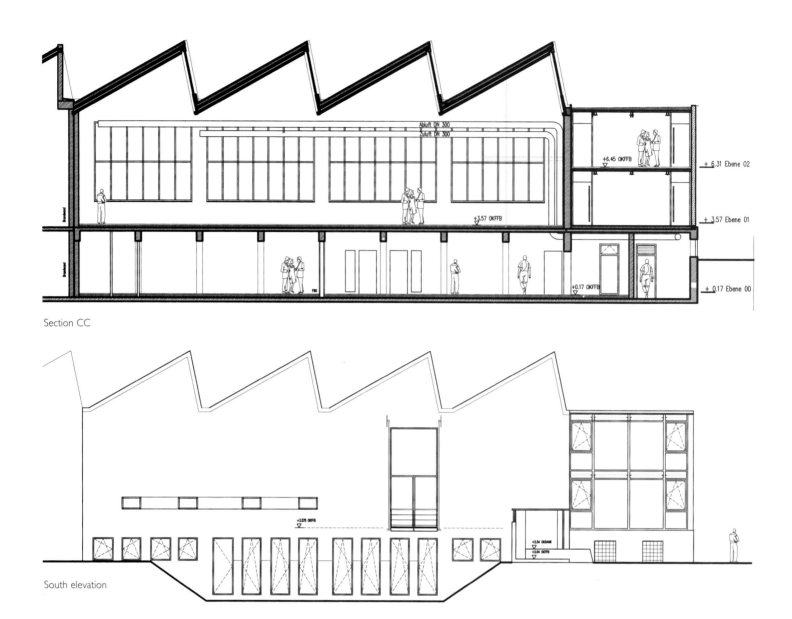

Section CC

South elevation

The ground floor houses the model workshop, a rest area, the kitchen, the terrace, the services and the architect's loft. From this level a megalith rises to the first floor, where the access to the building, the cafeteria and a studio are located. From the inside, the walls and roof of the building were insulated with panels of polyurethane, foam and mineral wool.

Section AA

Section BB

East elevation

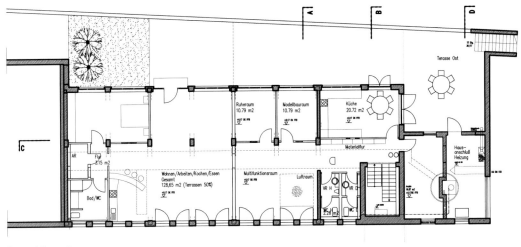

Ground floor plan

First floor plan

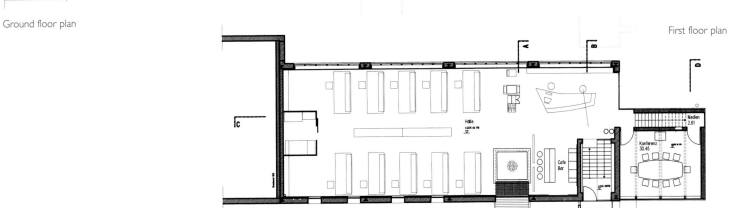

The scheme was inspired by a desire to create an open, flexible working environment that favored communication between the workers. Thus, the studio occupies a large transparent space with very high ceilings and industrial overtones. The desks are on wheels, which facilitates mobility. The upper floor also houses a meeting room.

An extension was added to the two floors of the original building. The addition to the shed hall was built using wooden studs and features a stainless-steel facade. The exterior of the building is in brickwork.

Ignacio Mendaro Corsini
San Marcos Cultural Center and Toledo Municipal Archives

Toledo, Spain

The project was made up of three clear and distinct, yet interconnected, parts. The first was the consolidation of the church ruins to keep it from collapsing, the second was its rehabilitation and adaptation for its new use as a cultural center and the third was the adding on of a new floor to house the Municipal Archives of Toledo. Each phase had to be linked to the others, forming part of the whole with the creation of private patios and a public square.

When building the Municipal Archives, which was to occupy part of the plot where the old convent once stood, the architects took on the commitment to reconstruct an urban fabric whose public face was the most architecturally degraded portion of the complex of buildings.

An immense plinth-like wall was built which follows the foundation lines of the old convent and which brings out the true dignity of the existing building: the monumental volumetry of the church with its lateral naves and sacristy.

The strength of wall engineering, so commonplace and deeply rooted in our cities, was sought. The wall was built of concrete and, in spite of the voices raised against a project of such characteristics in a historically significant city, the design scheme refers precisely to the timelessness of this material which the Romans and Arabs used long ago and which is still in current usage. Large voids were opened up in this wall which, in their own way, further highlight the transparency of the inner patios.

Special care was taken with the planks of the formwork, which is, in places, done with thick steel plates, placed in a seemingly random formation, creating voids and lamps.

During the course of the work, the discovery of archaeological remains obliged changes in the design scheme, thereby turning a problem into the project's virtue – the remains of the past now live alongside the new architecture. Thus, the entranceway via the archaeological remains has been given a minimalist treatment in the effort to give a natural response to a cultural requirement; modifications have been made to the Archives whenever a Roman furnace or Arabian well was found. Far from covering them up, they have been put on display, adjusting forms, lighting and architectonic resources accordingly.

Photographs: Lluís Casals

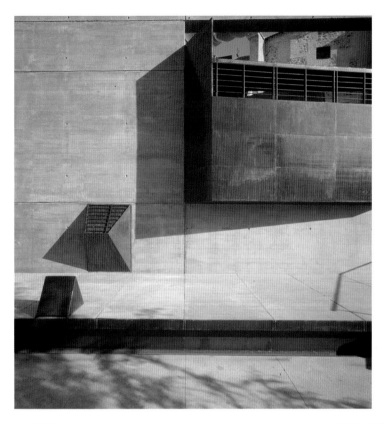

231

The concrete wall of the Municipal Archives has been brought into harmony with the golden hues of
Toledo through the use of natural coloring mixed into the mortar.

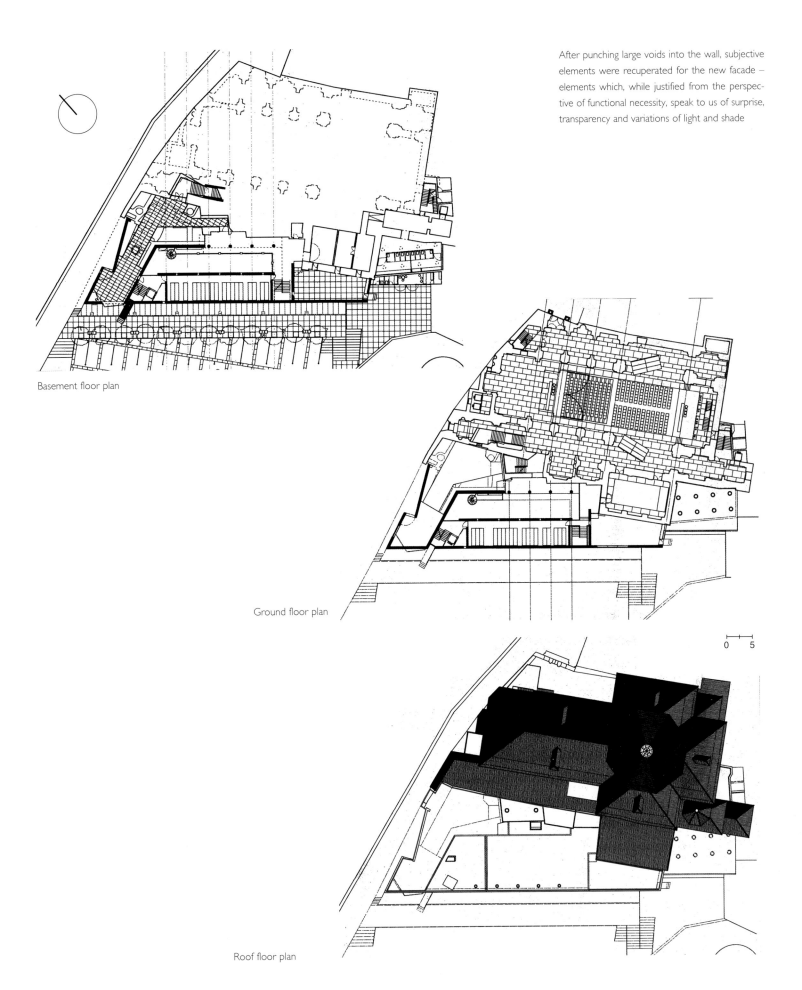

After punching large voids into the wall, subjective elements were recuperated for the new facade — elements which, while justified from the perspective of functional necessity, speak to us of surprise, transparency and variations of light and shade

Basement floor plan

Ground floor plan

0   5

Roof floor plan

233

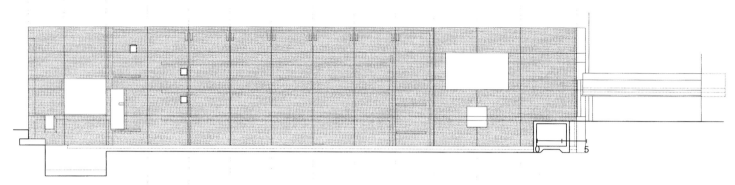

Southern elevation

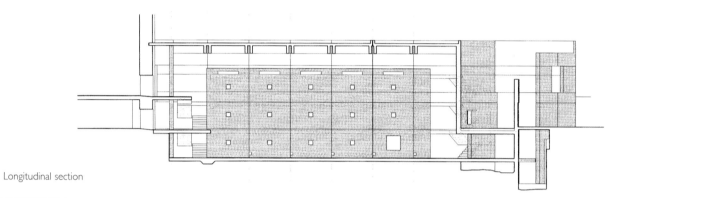

Longitudinal section

Cross section

0   5

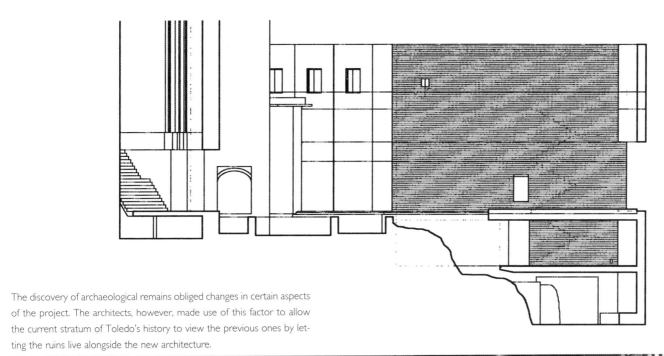

The discovery of archaeological remains obliged changes in certain aspects of the project. The architects, however, made use of this factor to allow the current stratum of Toledo's history to view the previous ones by letting the ruins live alongside the new architecture.

Manuel de las Casas
Hispano-Luso Institute
"Rei Alfonso Henriques"

Zamora, Spain

The scheme for the Institute of Spanish and Portuguese Studies, built within Gothic ruins, was simple: to enhance the beauty of the incomplete –the evocation of a past era– and to design a program with minimalist volumes.

A Z-shaped building, which delimits the church's former void, divides the space in two: a public garden formerly occupied by the church's three naves and another on the spot where the old convent's first cloister once stood.

A historical analogy thereby arises: a public space dominated by the classrooms and library, recalling the chapels and choir, and a private space where the dormitory rooms are located – some within the volume of the new floor, and others occupying the ruins of the existing nave on the southern portion of the plot.

As the nexus between the new and existing structures, the roof soars over the built body to meet the church's perpendicular nave, whose limits begin with the Chapel of Escalante, giving rise to a wide porch. This roof shelters the vaulted Great Hall, the old storeroom which will be used for ceremonies and events, as well as a cafeteria placed above this hall.

The architectural remains have been occupied, thus making manifest the original idea of enhancing the value of the ruins. The chapel (Capilla de San Buenaventura) abutting the entrance has been restored, in order to make use of it as a reception hall, and the apse has been converted into an entrance portico.

A room has been built within the Chapel of the Dean, which will be used as an exhibition and conference room. Its "floating" roof is an incomplete rectangle, dramatically bringing light into the interior and metaphorically recuperating the Hall's original volume. The entrance to the grounds is via the oldest door at the head of the transept, thereby creating a tangential access to the ruins so that visitors see the apse only once inside.

Photographs: Ángel Baltanás & Eduardo Sánchez

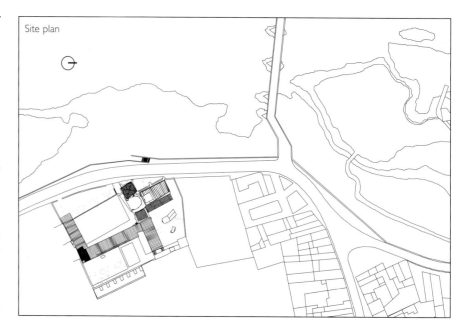

Site plan

First floor plan

Second floor plan

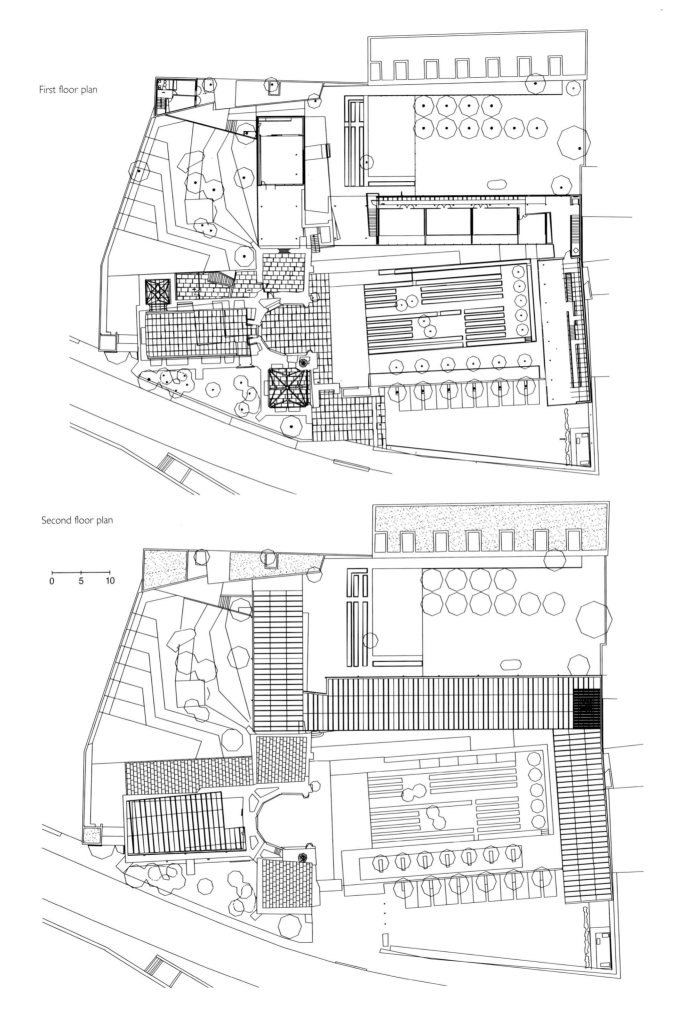

0   5   10

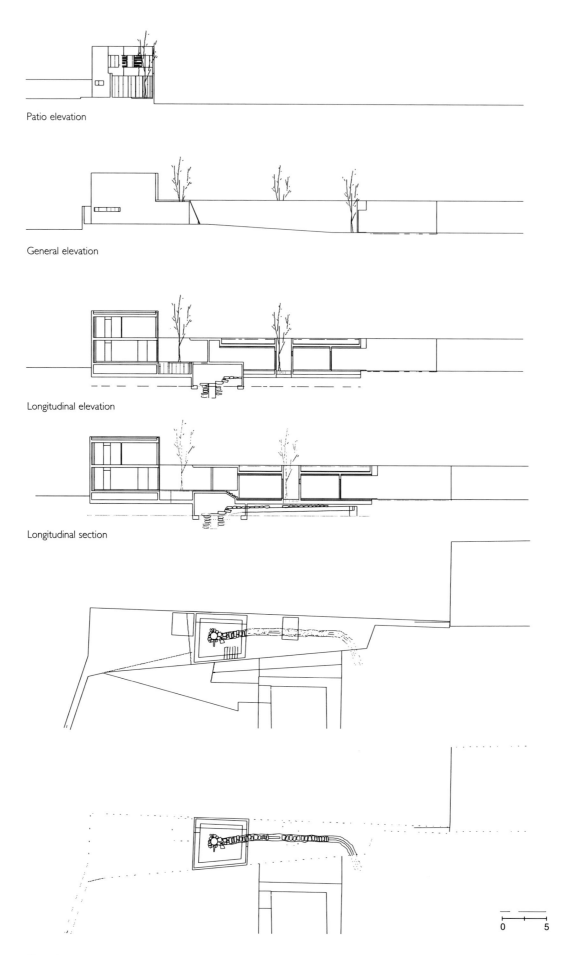

Patio elevation

General elevation

Longitudinal elevation

Longitudinal section

0   5

Floor plans

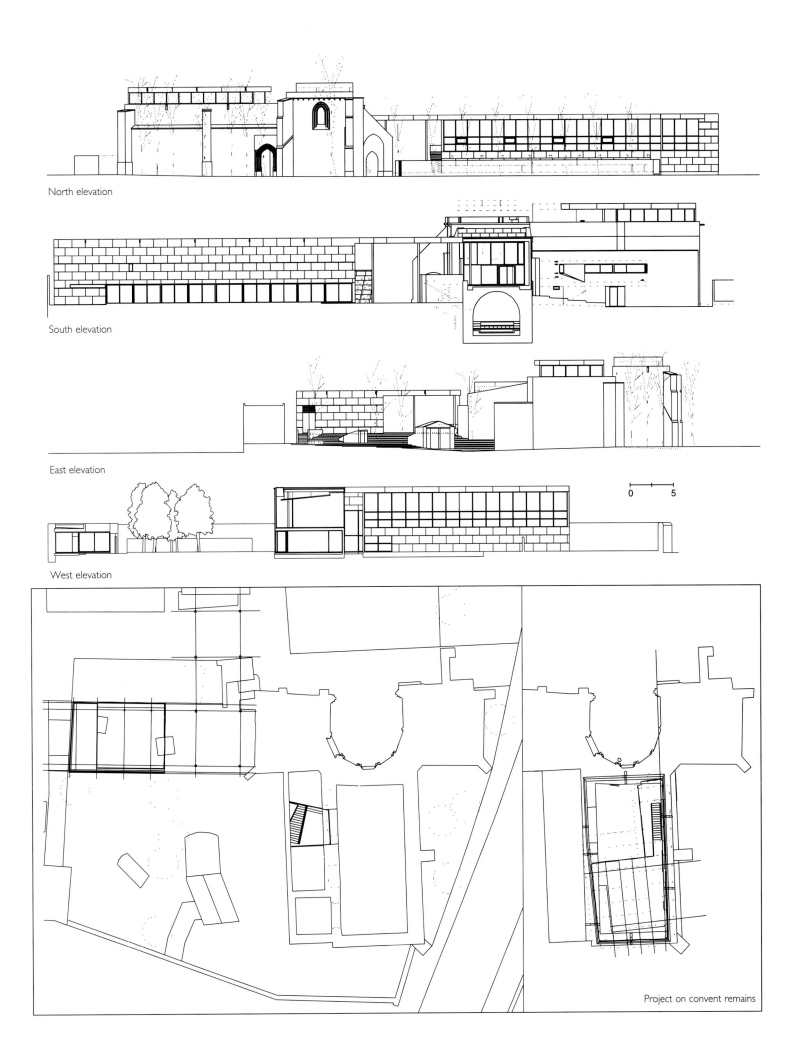

North elevation

South elevation

East elevation

West elevation

0 — 5

Project on convent remains

245

Sections

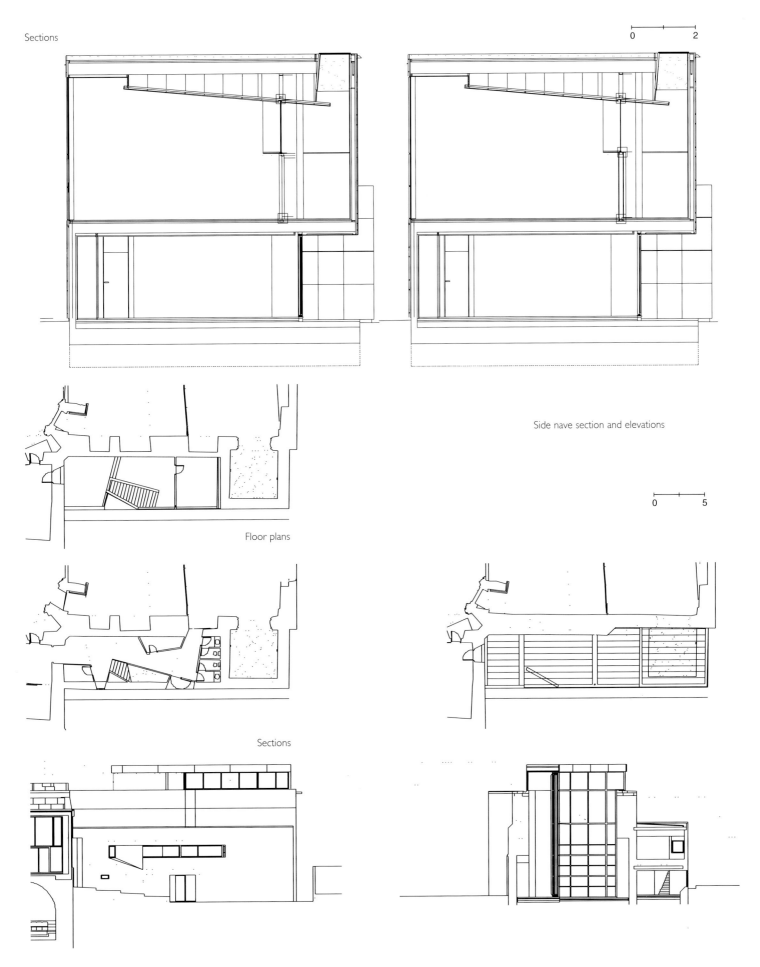

0   2

Side nave section and elevations

0   5

Floor plans

Sections

247

Elevation and roof sections

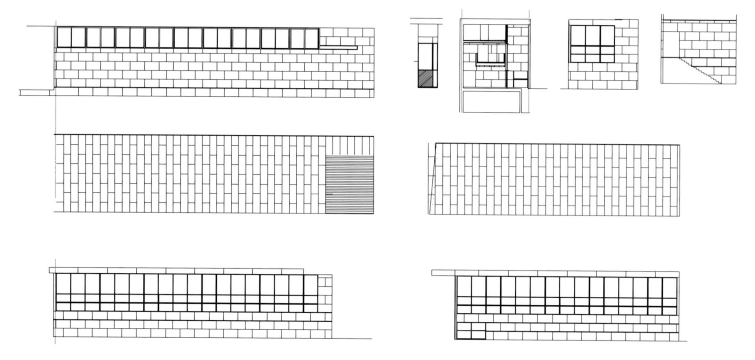

249

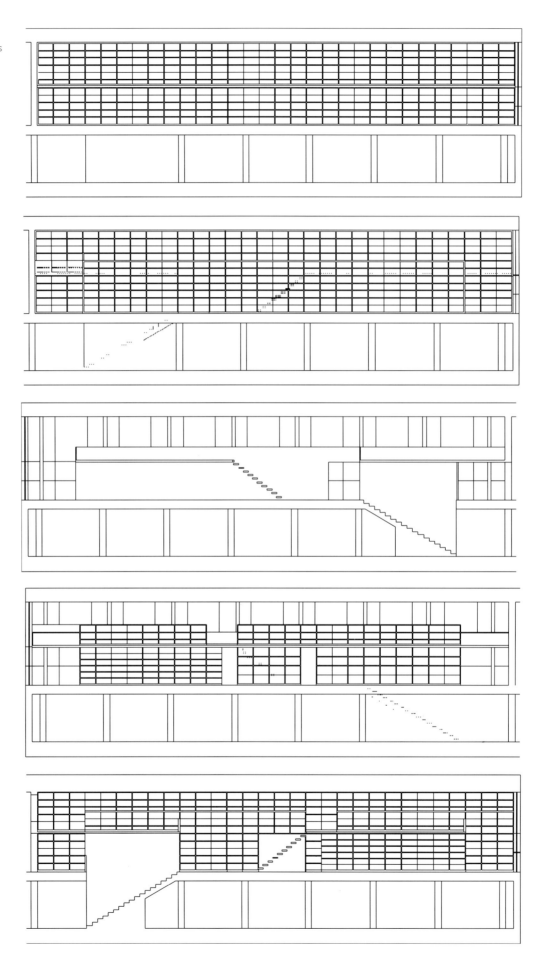

Library floor plans

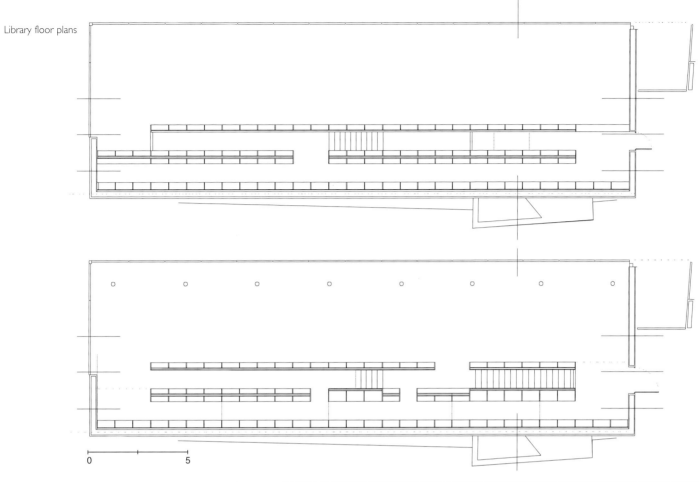

0   5

Prof. Jürg Steiner
The Coking Plant /
Exhibition space

Essen, Germany

In the conversion to an exhibition site, the Coking Plant had to be equipped especially with "traffic areas": rails, stairs, bridges and additional floors. The parts of the plant to be viewed are the weighing tower, the mixing plant and the 140-meter-long conveyer bridge. The exhibit Sun, Moon and Stars has been set up throughout these spaces and is conceived as a covered tour.

The main exhibition building is the 35-meter-high mixing plant, to which a fourth floor has been added to the existing three. Two thirds of the cubic volume were occupied by twelve bunkers with high, windowless walls of raw concrete with a strip-like skylight. Sixteen large openings now link the bunkers, transforming them into galleries.

Strip windows have been installed wherever necessary. The long sides of the building were without vertical intermediate pillars and could therefore be fitted with horizontal lighting slits along the entire length of the building. The structure that is visible from the outside is merely a cladding and is only twelve centimeters thick.

The conveyer bridges, which were originally glazed on both sides in their upper half, were later clad in corrugated steel. The conveyer bridge between the weighing tower and the mixing plant has subsequently been freed from its ugly armor and given a new upper part, which is now less framed and slightly higher than the original.

The old elevator shaft in the former main stairwell, which is now the fire escape route, has been fitted with two additional landings to serve the newly created floor on the bunker level and the reception point on the roof.

To do this, it was necessary to raise the height of the staircase tower to enable visitors to reach the upper landing.

A 140-meter-long cable system with four cabins conveys the public from the weighing tower to the top floor of the mixing plant.

In the conversion process, the coke oven batteries –technically the plant's central elements– were not overlooked. Battery 9 has been broken through along its length, providing an open view of its inner workings.

Photographs: Steiner Architectural Office, Werner J. Hannappel,
Frank Vinken & Joachim Schummacher

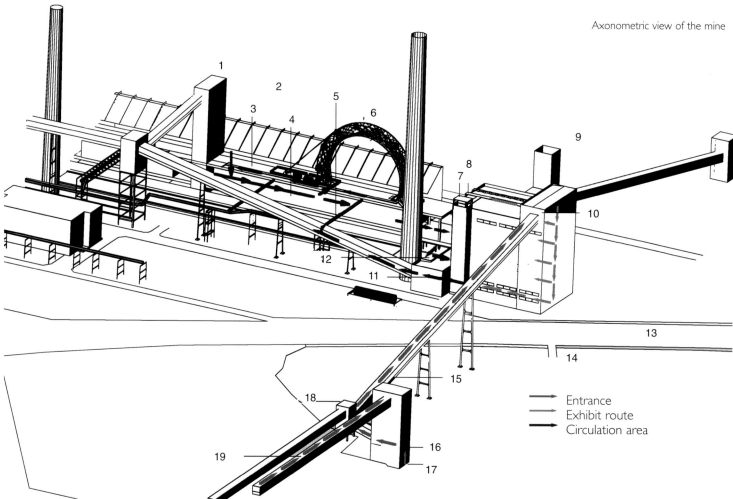

Axonometric view of the mine

Entrance
Exhibit route
Circulation area

1. Intermediate bunker
2. Battery 9
3. Field of sunflowers
4. Ovens
5. Walkway over the pipe bridge linking Battery 9 to the conveyer bridge
6. Ferris wheel

7. Uppermost point of elevator
8. Viewing platform
9. Mixing plant
10. Upper station
11. Coat check, exit, exhibit shop, cafeteria
12. Walkway between the chimney stack and mixing plant

13. Lawn
14. Parking lot
15. Bridge with two trams
16. Ticket office, coat check, entrance
17. Entrance to the old weighing tower
18. Lower station
19. Entrance to the XII Zollverein mine

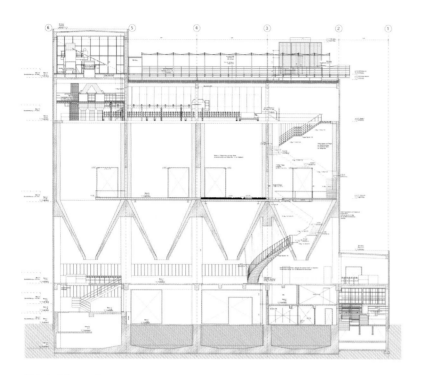
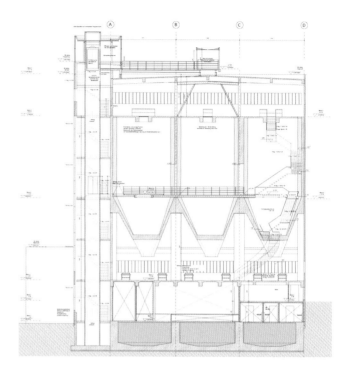

Main building sections

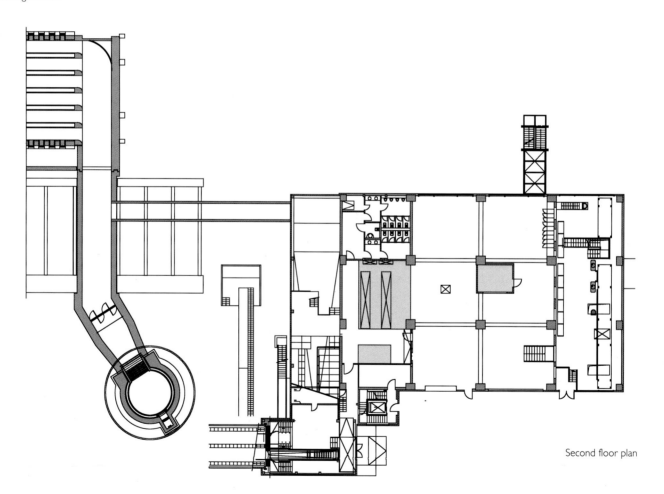

Second floor plan

Since the main load-bearing structures were placed throughout the space, instead of being concentrated along the facades, it was possible to install glazed surfaces along the perimeter.
As part of the plant's conversion into an exhibit space, bridges and walkways were added to facilitate circulation with wheelchairs and prams.

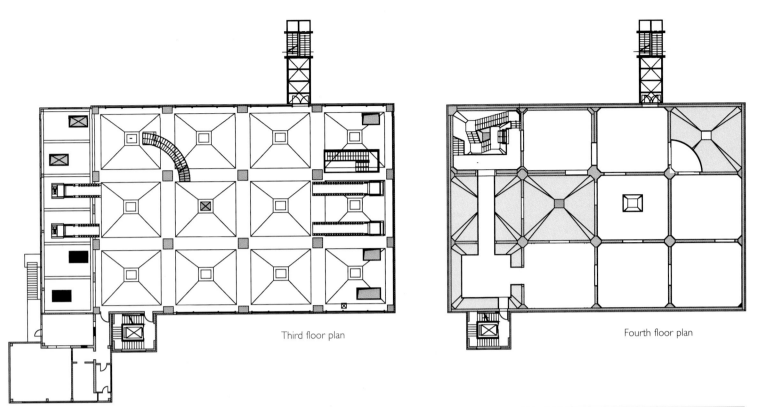

Third floor plan

Fourth floor plan

Although the architectural language of the plant was determined by functional demands, the renovation process has highlighted its high artistic value, as seen in the incorporation of the old coking ovens into the exhibit space. Two new landings have been added to the massive concrete-clad stairwell, which was built from the old elevator shaft.

Fifth floor plan

Sixth floor plan

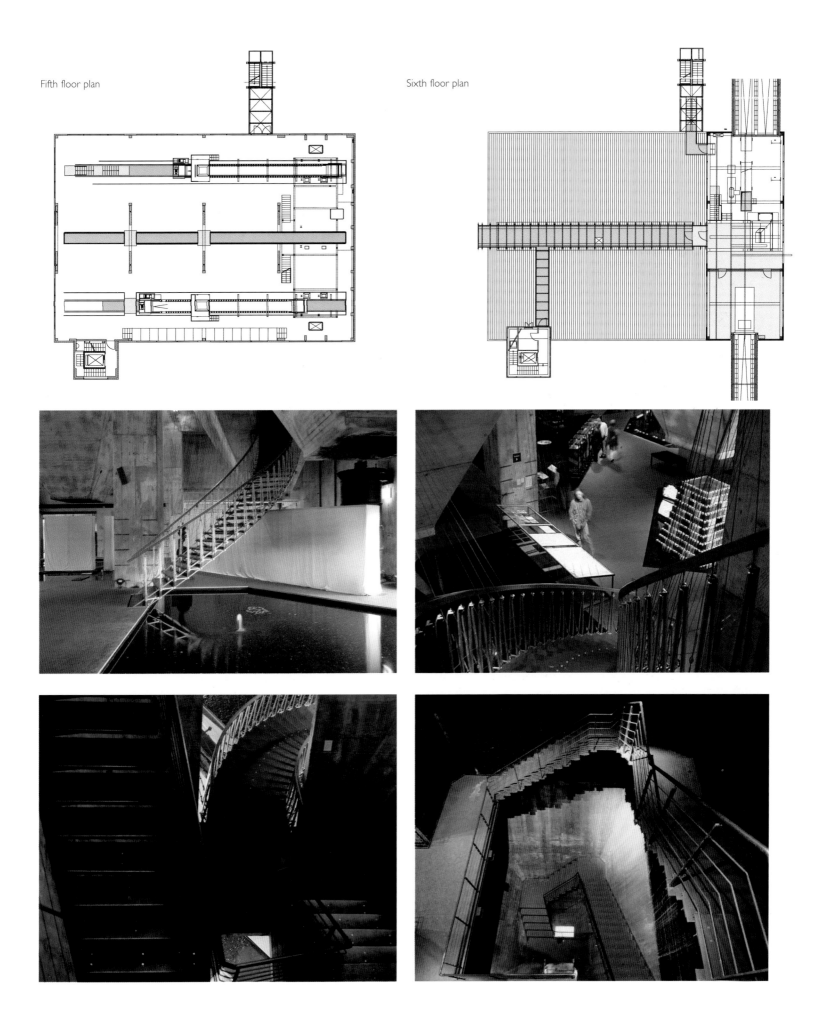

Benoîte Doazan & Stéphane Hirschberger, architectes
Rehabilitation of a Covered Market

Lagny-sur-Marne, France

The rehabilitation of this unique building —foods market on the ground floor and library on the first— called for improving the facade, installations and services. Since the market was to be undergoing reforms, the opportunity presented itself to give both volumes a similar treatment.

The program included changing the cladding on the top floor for a new facade of untreated red cedar clapboard affixed to the market pillars, thereby emphasizing their rhythm. Wood was chosen because it is lightweight, durable and easy to affix to the building. These prefabricated facade panels shield the building from water and wind, while at the same time giving coherent organization to the library and allowing greater freedom for interior work. The cladding profile is a cantilevered copper cornice resting on wooden modillions.

The lower part of the facade (where the market is located) consists of sectional fiber glass doors which run from pillar to pillar and can be folded up to fit inside a compartment hidden behind a decorative panel. At night, their translucence gives the building the look of a lamp on which are projected shadows of shopkeepers and customers.

The pillars inside the market are painted, except at the base, where concrete skirting has been added for greater resistance.

The two floors are divided along the facade by an armor-plate glass canopy, which recalls familiar images of markets from times past and enables the stalls to be extended to the edge of the built space.

A small square —ideal for taking a break or socializing— was opened up in the center of the market as a result of the reorganization. The stalls lie parallel to the slope and enjoy fairly uniform distribution, while the cross aisles break up the stalls into three commercially viable units. In an attempt to open the market out toward the town and, likewise, draw the surroundings inside, the floor paving is the same as that of the street.

Photographs: Atelier Doazan-Hirschberger / Jean-Marie Monthiers

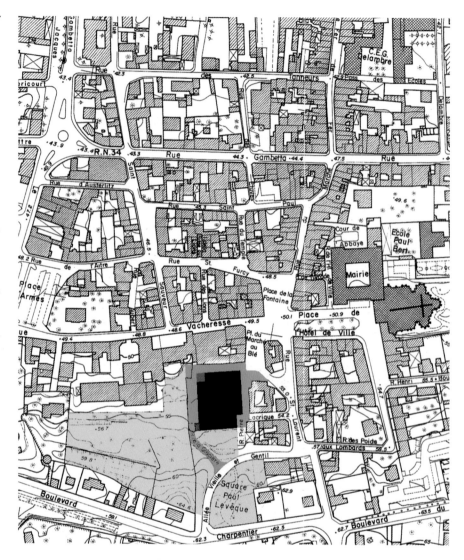

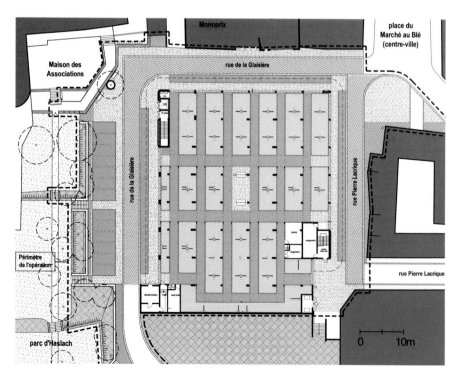

261

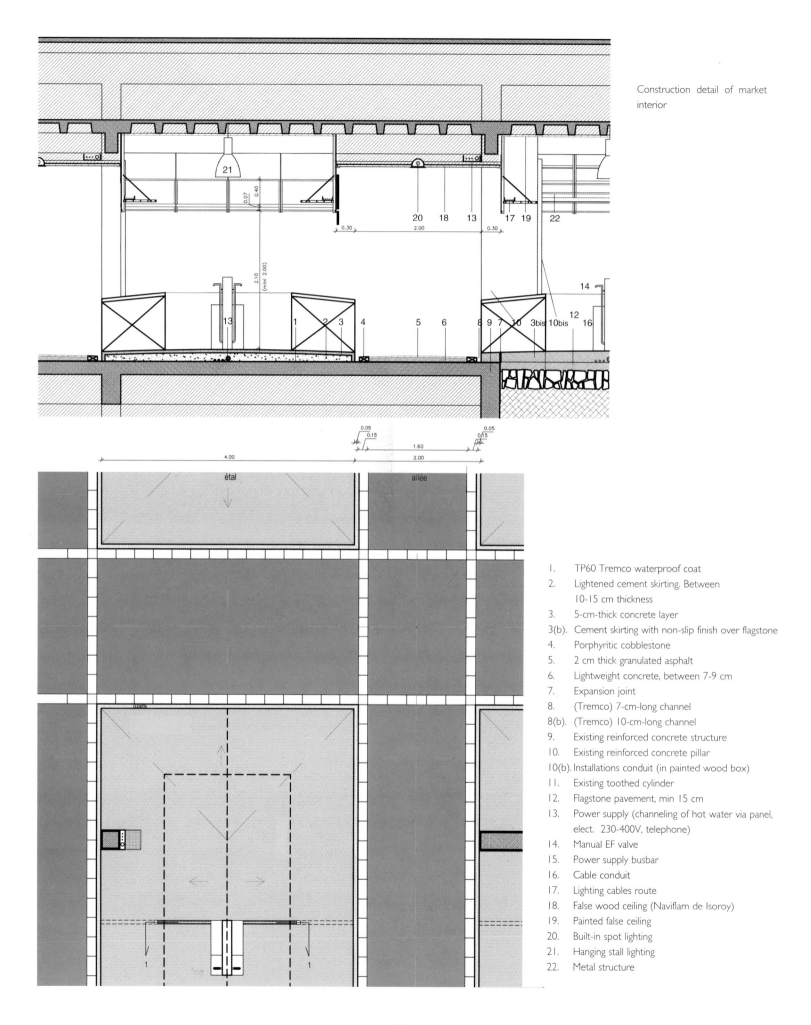

Construction detail of market interior

1. TP60 Tremco waterproof coat
2. Lightened cement skirting. Between 10-15 cm thickness
3. 5-cm-thick concrete layer
3(b). Cement skirting with non-slip finish over flagstone
4. Porphyritic cobblestone
5. 2 cm thick granulated asphalt
6. Lightweight concrete, between 7-9 cm
7. Expansion joint
8. (Tremco) 7-cm-long channel
8(b). (Tremco) 10-cm-long channel
9. Existing reinforced concrete structure
10. Existing reinforced concrete pillar
10(b). Installations conduit (in painted wood box)
11. Existing toothed cylinder
12. Flagstone pavement, min 15 cm
13. Power supply (channeling of hot water via panel, elect. 230-400V, telephone)
14. Manual EF valve
15. Power supply busbar
16. Cable conduit
17. Lighting cables route
18. False wood ceiling (Naviflam de Isoroy)
19. Painted false ceiling
20. Built-in spot lighting
21. Hanging stall lighting
22. Metal structure

Detail of facade construction

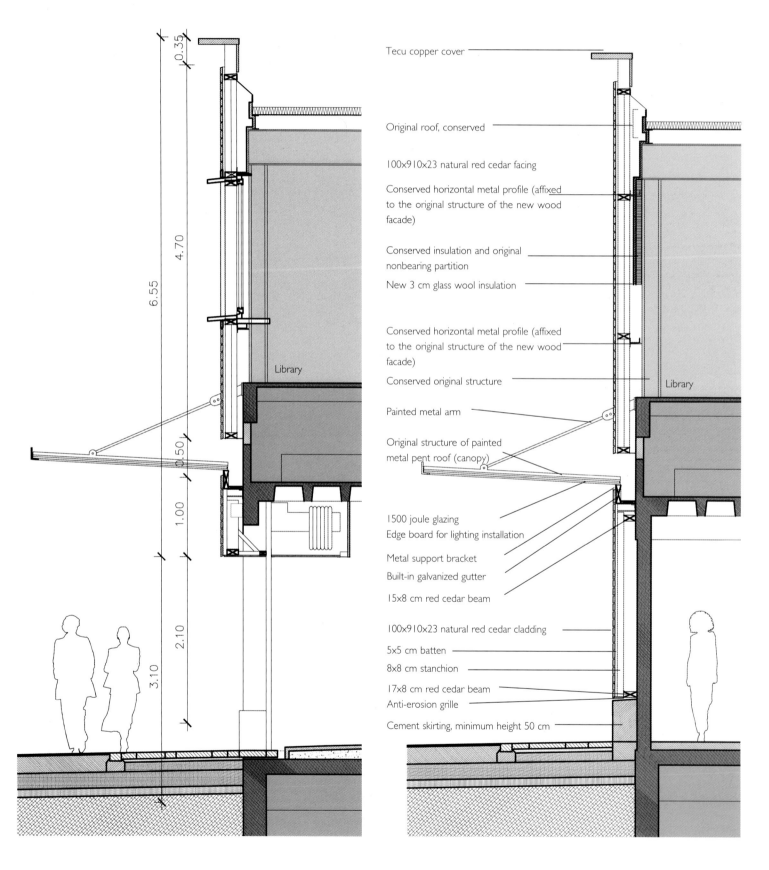

Tecu copper cover

Original roof, conserved

100x910x23 natural red cedar facing

Conserved horizontal metal profile (affi<u>x</u>ed
to the original structure of the new wood
facade)

Conserved insulation and original
nonbearing partition

New 3 cm glass wool insulation

Conserved horizontal metal profile (affixed
to the original structure of the new wood
facade)

Conserved original structure

Library

Painted metal arm

Original structure of painted
metal pent roof (canopy)

1500 joule glazing
Edge board for lighting installation

Metal support bracket
Built-in galvanized gutter

15x8 cm red cedar beam

100x910x23 natural red cedar cladding

5x5 cm batten

8x8 cm stanchion

17x8 cm red cedar beam
Anti-erosion grille

Cement skirting, minimum height 50 cm

0.35
4.70
6.55
0.50
1.00
2.10
3.10

Library

1. Copper cover
2. Natural red cedar cladding, structure and joinery in aged wood
3. Zinc cover
4. Painted cantilever, translucent glass
5. Translucent fiber glass doors
6. Painted concrete pillar

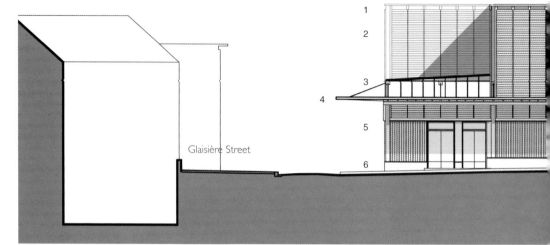

Glaisière Street

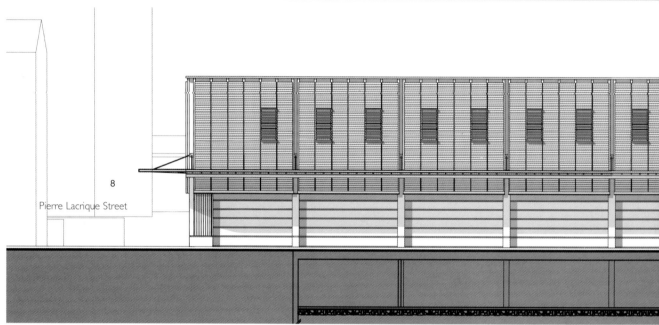

Pierre Lacrique Street

8

0   5   10 m

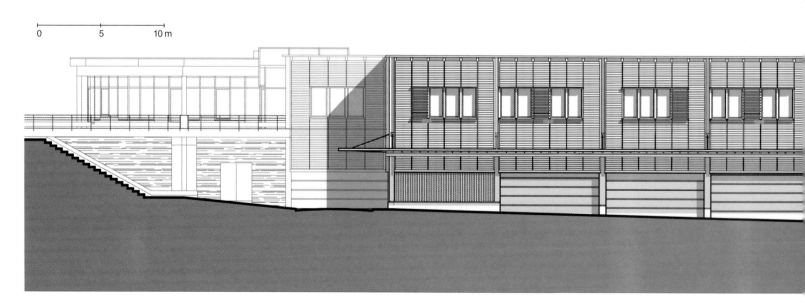

West elevation

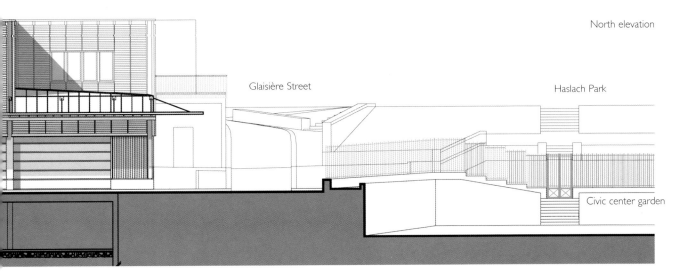

North elevation

Glaisière Street

Haslach Park

Civic center garden

East elevation

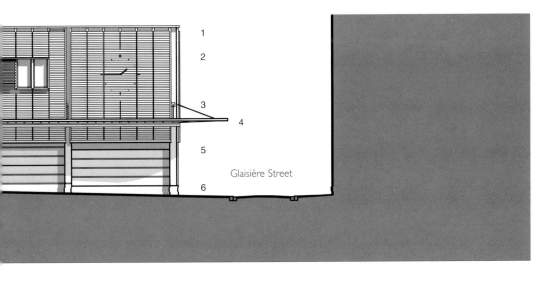

1

2

3

4

5

Glaisière Street

6

After the rehabilitation, the south face of the library now opens out toward a garden, while the market faces north, toward the city center.

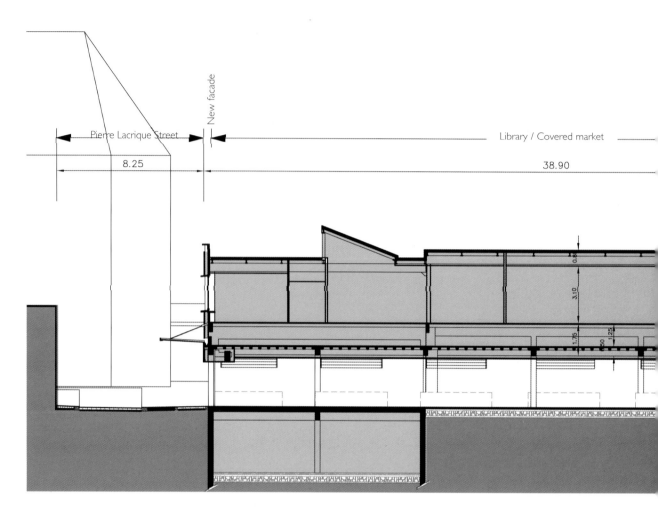

Pierre Lacrique Street

New facade

Library / Covered market

8.25

38.90

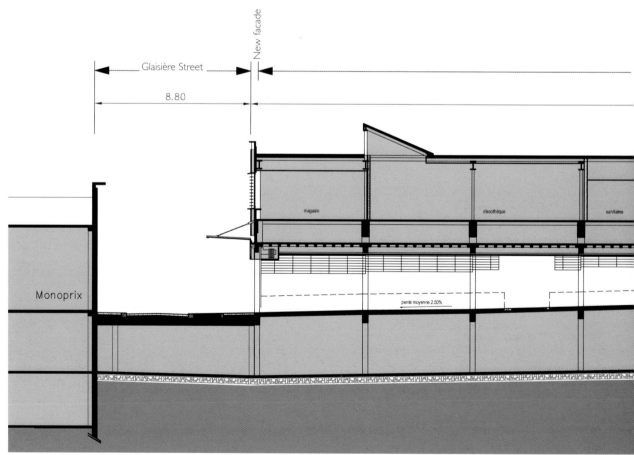

Glaisière Street

New facade

8.80

Monoprix

magasin

discothèque

sanitaires

pente moyenne 2.50%

North section

New façade

Glaisière Street

18.84

Haslach Park

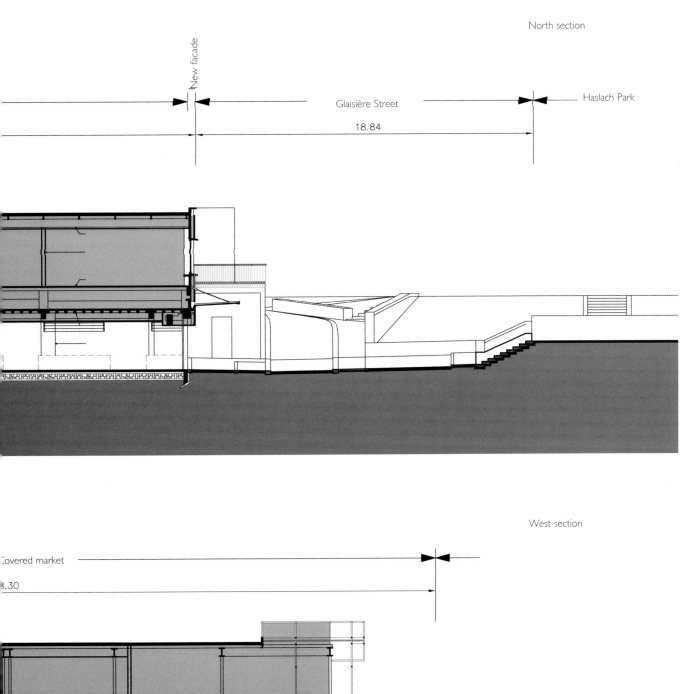

West section

Covered market

8.30

salle polyvalente   hall d'accueil   sas d'entrée

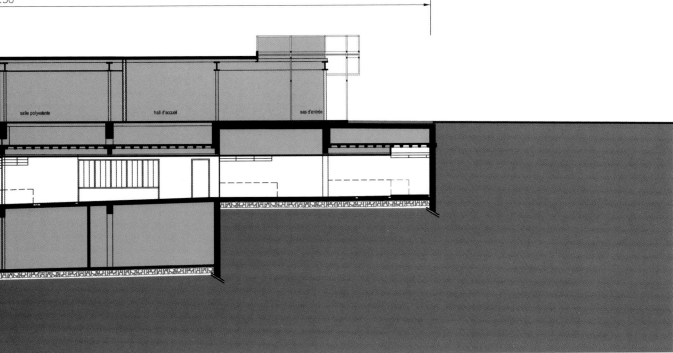

Correa + Estévez , arquitectos
(Maribel Correa y Diego Estévez)
Rehabilitation of the
Cabrera Pinto Institute as a
Museum and Cultural Center

Tenerife, Canary Islands, Spain

The Cabrera Pinto Institute is one of the most notable buildings in the historic and artistic district of the city of La Laguna. The structure dates from the beginning of the 16th century and was originally a convent and church. It consists of a main cloister, which is flanked by halls and connected to the Church of San Agustín. In the 18th century, a second cloister was built, adjoined to the original and to the church itself. Towards the end of the century, the grounds were being used as an educational institution, a function which has continued uninterrupted to the present. Its conversion into a place of learning necessitated a series of changes over the years, which ultimately adversely affected the structures, to the point where their original appearance was all but indiscernible.

The aim of the project was to completely rehabilitate the building, adapting it to the requirements of a museum and cultural center located in the oldest and most symbolic section of the building complex. Here, various rooms have been equipped for exhibits, ceremonies, a library and a newspaper archive. These new public spaces function independently of the center, while they are at the same time linked to its educational aspect. The administrative and teaching facilities are in the newest wing of the building.

The project called for "cleaning", tearing down the numerous superfluous structures in order to create large areas which would pave the way for a new museum. The Renaissance-era cloister, which was sitting in a partially ruinous state, was restored with preventive shoring on all of the structural elements. To do so, the garden had to be drained and the foundations of the perimeter wall, which itself was hardly intact, were strengthened. The patio galleries were also shored up and stripped down for the restoration of the elements of value, such as the double colonnade of red stonework, the roof and wooden lagging and ribs.

The original geometry of the second cloister floor plan was recovered through a process of tearing down all undesired components, individually analyzing the elements to be conserved and adding new enclosures and finishes.

The new classroom wing has been expanded, formalizing a third patio, which is connected to the other two, in the institutional area of the most recent construction. This creates a curious succession of cloister patios built at different points in the building's history.

Photographs: Carlos Anglés

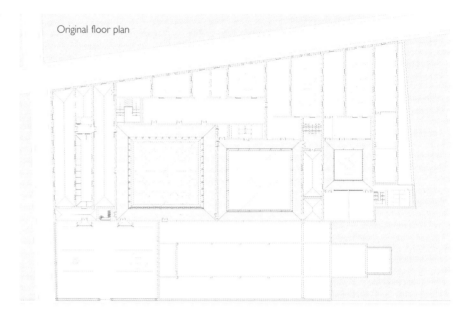

Original floor plan

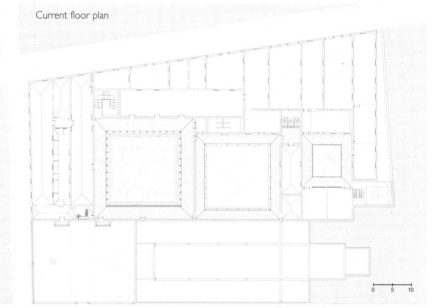

Current floor plan

A clean expanse of glass running the perimeter of the upper gallery is supported by a wood frame, which is fastened to the glass along the bottom edge; bronze pincer-like elements hold the glass in place along the upper arris. This separation allows ventilation at all times and also keeps out rain water, which is the primary cause of the deterioration of the upper gallery and its wooden fixtures.

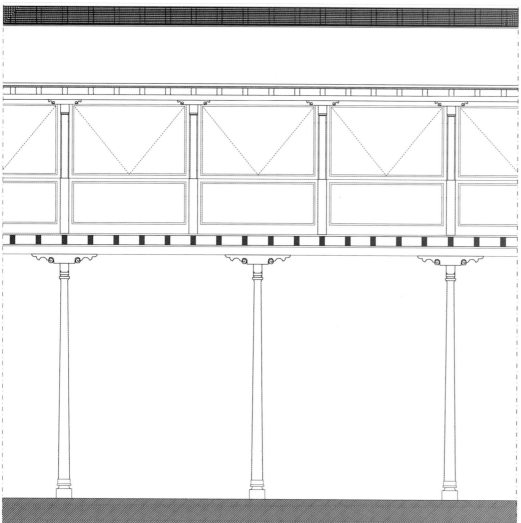

The wooden pillars of the upper gallery in the second cloister have been restored. A fringe of glazing has been installed below the eaves of the gallery roof, bathing the church's stone face in light and emphasizing its vertical continuity.

Four walled-in spaces from the 16th century settlement were discovered in one of the rooms. These have been formally recuperated with modern window carpentry.

Guido Canali
Water Mill

Modena, Italy

The object of the reform work was an old building whose nucleus dates from 1558. It had been used for productive functions, such as carpentry, until 1896, when it was turned into a mill.

The base of the floor plan is a rectangle, the length of which runs alongside the canal and is divided into three areas.

The main, northernmost space, where production took place, has a brick body and common pitched roof (corresponding to the mill). The grinding block is in the basement and, on the first floor, there is a large grain storage room with wooden trusses overhead.

Fortunately, the space which once housed the grinding block was in fairly good condition when it was acquired by the current owners – even the wooden scoops and millstones were still intact. The rest of the building, on the other hand, showed signs of having been subjected to a number of alterations.

The idea behind the project was to leave the millstone room intact, carefully restoring it while highlighting the theme of water, which had once been inseparably linked to the functioning of the mill. Thus, a glass bulkhead was installed, allowing observation of the interior of the waterworks and the scoop mechanism, both of which have been completely restored.

Adhering to client demands, a large apartment (1000 m^2) was created in the northern body and two independent apartments (104 m^2 and 130 m^2) were installed in the southern volume, plus a service room of 40 m^2. The entrances to the master bedrooms are located above the canal at the middle of the building. Natural light filters through the skylight and is reflected off the swimming pool in the basement. Metal walkways cross the void of the foyer on the different floors, linking the building's two wings.

In the central volume, the garden has been turned into a kind of room/porch. Large sliding windows set behind a portion of the old wall allow this space to be either open or closed, depending on the season.

Throughout the scheme, particular attention was paid to the use of traditional materials for the indispensable integration of the old factory.

Structural calculation: Francesco Canali
Collaborator: Angela Cacopardo

Photographs: Paola De Pietri, Alberto Muciaccia,
            Stefano Botti & Francesco Castagna

The dining and living rooms are located on the top floor of the volume of the former granary. Two hand-worked skylights with stainless steel frames concealed between the existing beams illuminate the space.

Outside, water from the old canal has been channeled through a watertight underground conduit which runs under the garden, ending up at its origin on the other side of the house.

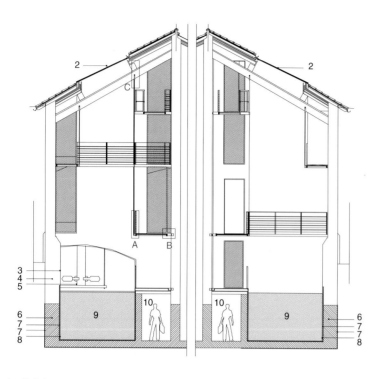

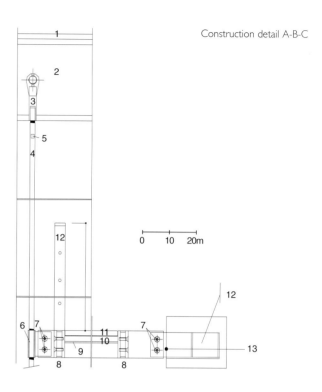

2. Skylight
3. Barefaced brick
4. Old brick wall
5. Old mill blades
6. Reinforced concrete foundations
7. Insulating casing
8. Custom-cut ceramic plate cladding
9. Swimming pool / Piscina
10. Swimming pool pump room

1. Wooden shore
2. Iron beam divided into three to affix ties
3. Stem to affix tie
4. Chain, threaded at both ends
5. Reductions to facilitate chain movement
6. Stem linking crossbars to chains
7. Channels, 10 mm diameter

8. Crossbar with two iron profiles with a 40x10 rectangular section, separated by 20 mm
9. 5-mm-thick plate
10. Flooring base
11. Flooring
12. Allowance for cross rigidity for the horizontal plate in load distribution
13. Edge of exposed wall

Cross section

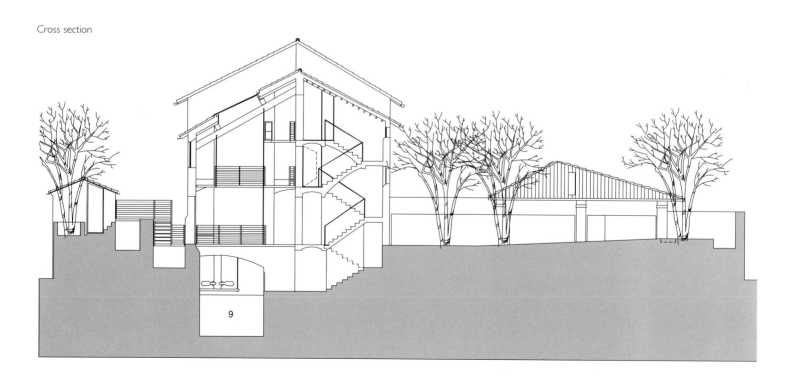

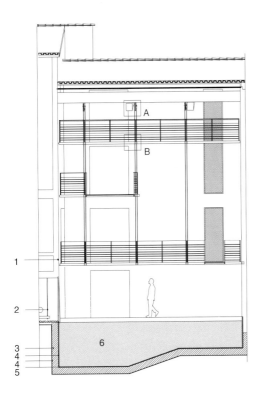

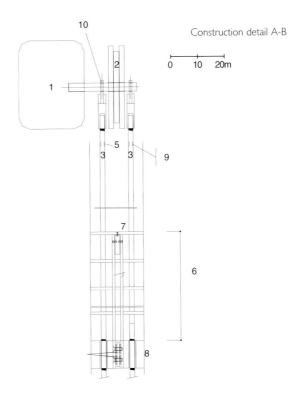

Construction detail A-B

0   10   20m

1. Barefaced brick
2. Old mill blades
3. Reinforced concrete foundations
4. Insulating casing
5. Cladding of custom-made ceramic plates that simulate barefaced brick
6. Swimming pool

1. Wooden shore
2. Iron beam divided into three for affixing ties
3. Chain, threaded at the ends
4. Reductions for facilitating chain movement
5. 10 mm channels
6. Post for handrail, comprised of two 40x10 mm profiles, separated by 20 mm

7. 40x10 mm handrail connected to the post separater
8. Washers to add thickness
9. Each of the reductions to be made in the chain area for their operation via a pin which cannot be larger than .4 cm.
10. Milling for affixing tie

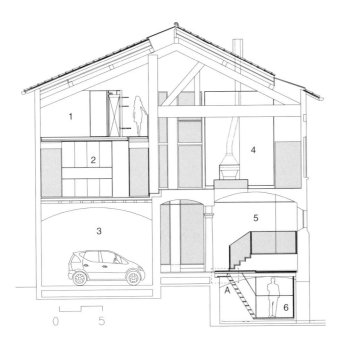

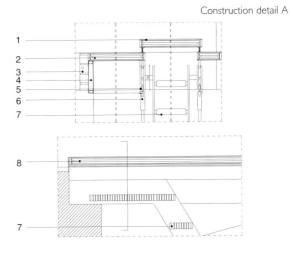

Construction detail A

1. Services
2. Dining room
3. Garage
4. Day area
5. Museum of old machinery
6. View of old mill-works

1. Movable bulkhead for accessing staircase
2. Large fixed window, with drawn out stainless steel fixtures
3. Half-brick cladding
4. Insulated loadbearing structure
5. Loadbearing beams
6. Staircase suspensions
7. Staircase in stainless grating
8. Movable, transparent glass bulkhead with stainless steel frame for accessing staircase

North facade elevation

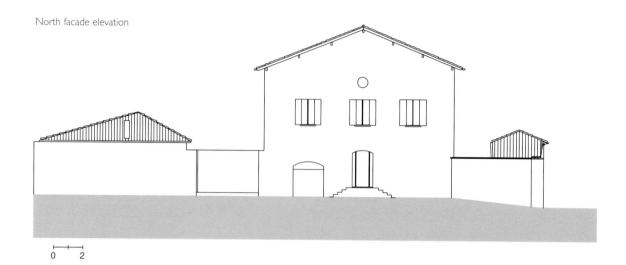

0   2

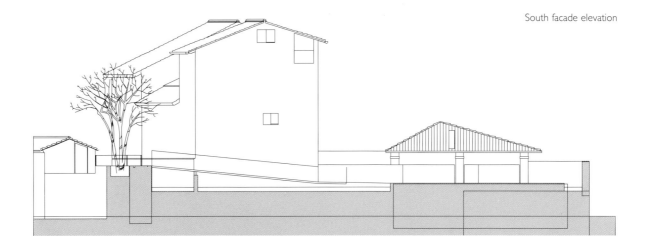

Basement floor plan

0 5

Ground floor plan

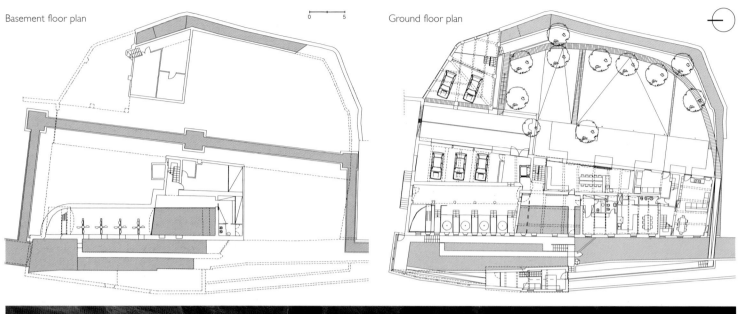

Stéphane Beel & Lieven Achtergael
Conversion of the Tack Tower into an Arts Production Center

Kortrijk, Belgium

While the structure of the Tack Tower and much of its surroundings have been respected, the proposed conversion of this industrial building into a production center for the arts necessarily entailed some changes.

The addition of a 3-meter-wide volume covers the entire width and height of one side of the Tack Tower and houses the new functions, while transcending the strictly functional. The modified appearance of the tower gives visual shape to the shift in use; and, at the same time, the essence of a 'production' tower is retained. The added volume contains the staircase, elevator, and toilet facilities, and serves as a vertical foyer, which ensures the autonomous use of the respective spaces.

The design scheme provides a whole range of possibilities to optimize the qualitative use of the building. Two sets of stairs form a circuit within the building. Spaces can be grouped; and no single area has been given a definitive use. This, combined with moveable furnishings, allows for a highly varied range of possible uses.

The facade of the Tack Tower oriented towards the city becomes a billboard (with a film projection screen on the top floor) and takes on a signal function. The old silhouette remains visible behind its alternately translucent and transparent skin.

The new volume on the south facade functions as a sunshade, while the more glazed north face provides a visually-interesting source of natural light. The top story combines a terrace, with a panoramic view of the center of Kortrijk, with an open-air film club. The roof covers part of the outdoor event area, serving as an awning oriented towards the renovated inner zone. The variety of conditions means that the new site will not just be a garden, but an attractive, functional, and public open-air building. The result is a complex of rooms "without walls but with a roof" and green, open-air chambers "without a roof but with walls".

Photographs: Jan Kempenaers

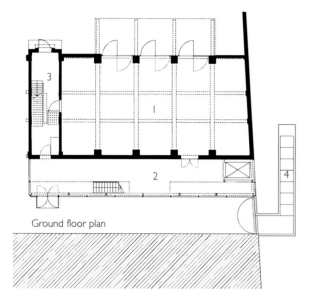

Ground floor plan

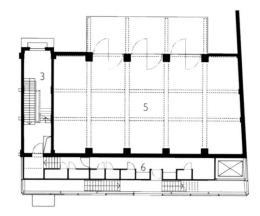

Intermediate floor

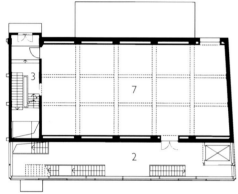

First floor plan

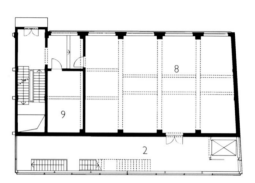

Second floor plan

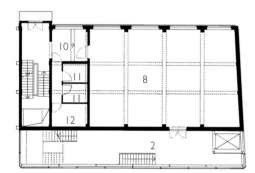

Third floor plan

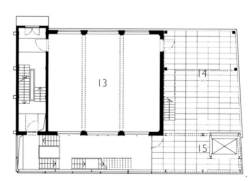

Fourth floor plan

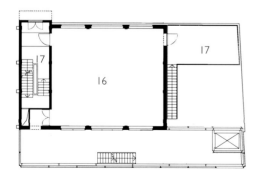

Fifth floor plan

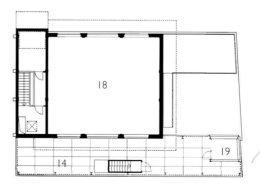

Sixth floor plan

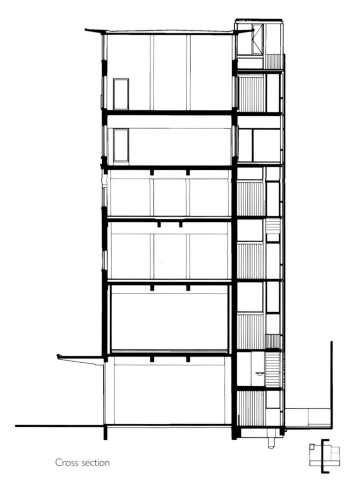

Cross section

294

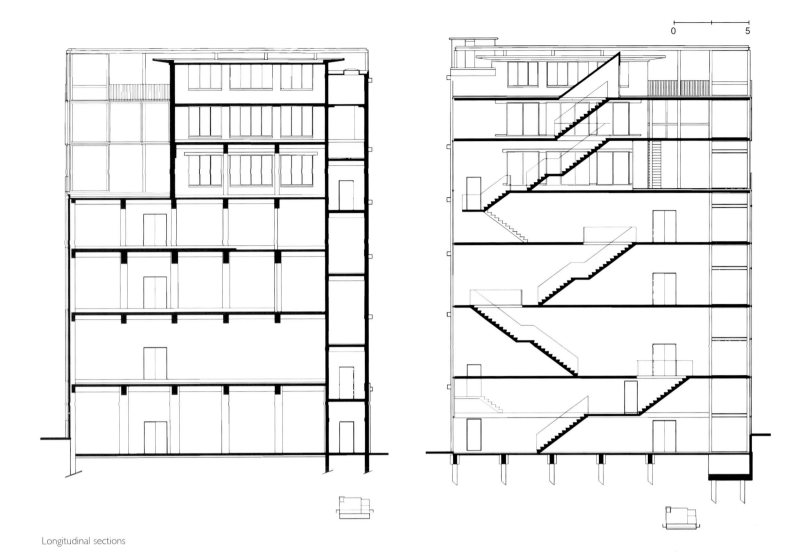

Longitudinal sections

Cristian Cirici & Carles Bassó
Vapor Llull

Barcelona, Spain

The Vapor Llull (a steam-driven factory), in an old industrial district of Barcelona, consisted of a set of buildings dating from the early 20th century which had been devoted to manufacturing chemical products.

The basic structure of the complex consisted of a long ground floor plus two floors, the highest of which had a sloping roof supported by a structure of wooden trusses. The complex also included a series of auxiliary premises adjoining the main building and a magnificent brick chimney measuring over thirty meters in height that was part of the steam engine that powered the factory. The architects decided to conserve the chimney in order to maintain a symbol of a time in which the whole district was full of steam-driven factories. The most suitable property for conversion into loft dwellings was the long main building. In order to create an open, private space and provide a one-vehicle parking space for each of the eighteen units into which the scheme was subdivided, a series of auxiliary buildings were demolished. To give independent access to each module of approximately 90 square meters, three sets of vertical communication elements were introduced, each with a stairwell and a panoramic elevator. Their formal expression gives the appearance of silos covered with enamelled corrugated steel. On the outside the main building was painted with silicate paint applied directly to the bricks, which were first stripped of their render.

In the interior, the spaces were left free and unfinished, so that each loft could be arranged according to the wishes of the different interior designers that were chosen to finish off the scheme. The layout and decor of this loft are by Inés Rodríguez. It is an two-level apartment in which a mezzanine houses the bedroom and a bathtub. It is a curious habitat in which the light and the space create an atmosphere of elegance.

Photographs: Rafael Vargas

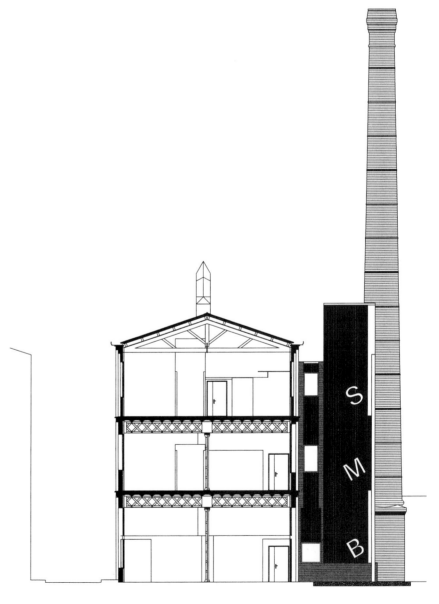

Ground floor plan

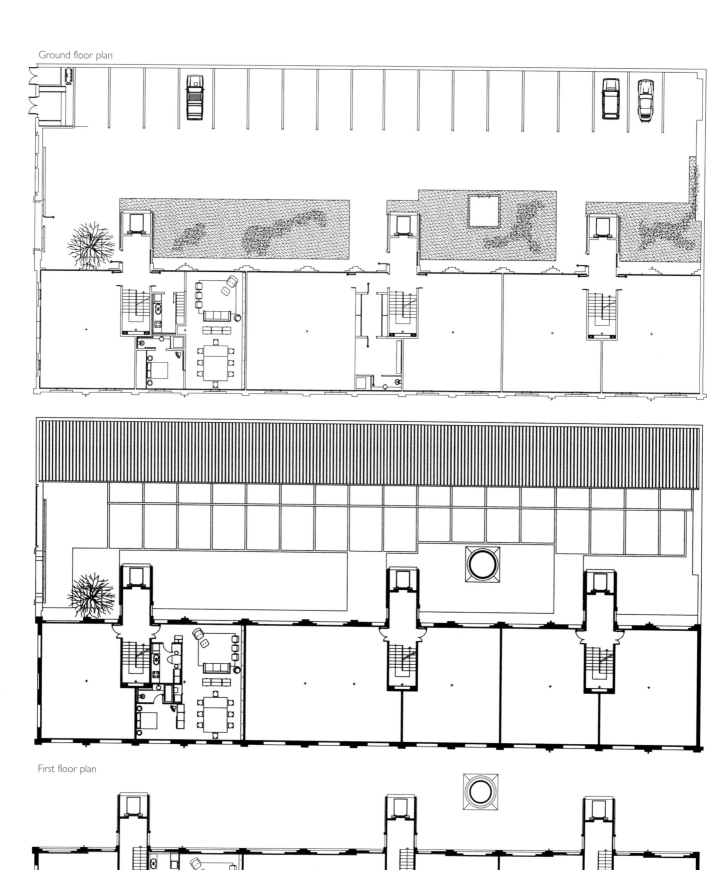

First floor plan

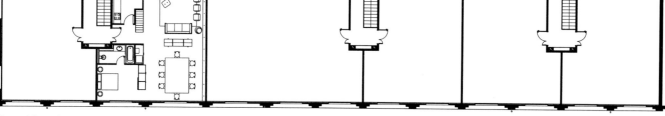

Second floor plan

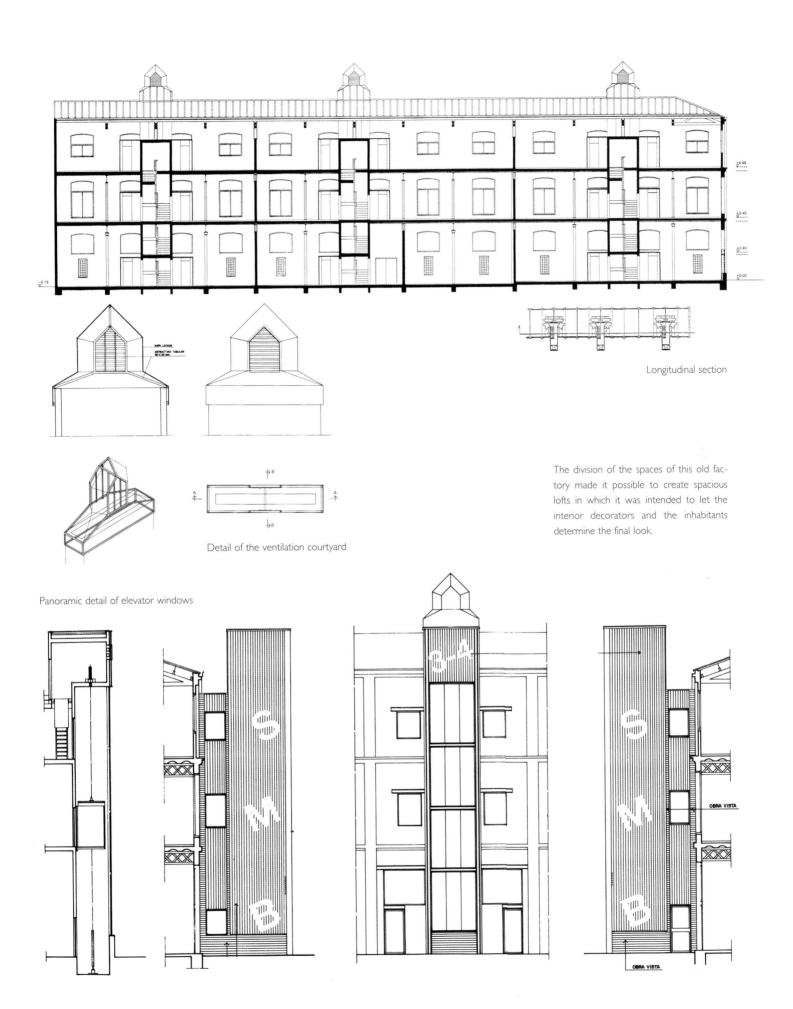

Longitudinal section

Detail of the ventilation courtyard

The division of the spaces of this old factory made it possible to create spacious lofts in which it was intended to let the interior decorators and the inhabitants determine the final look.

Panoramic detail of elevator windows

OBRA VISTA

OBRA VISTA

Large spaces with bright, clear walls predominate in this apartment. The most outstanding features at first sight are the wooden trusses in the whole dwelling and the polished concrete floor.

Aneta Bulant Kamenova
& Klaus Walizer
Conversion of Villa Sailer

Salzburg, Austria

The owners of this villa are art collectors who wanted to connect the house to the garden with spaces that should be protected from the weather and which could be used year-round. The brief was compatible with the architects' idea of using ecological building techniques to reduce climatic differences.

After eliminating all the unnecessary building elements that were added in the '50s, the form of a closed prism reappeared. On the north side, a garage for two cars and an entrance in the style of a small porch enclosed in satin-finish glass were added, both connected with an overhanging glass roof.

The main interest of the architects was to open the house toward the garden located to the south. To achieve this, they created a large terrace in which the glass-covered garden meets a pergola with tensed cables. The key to this spatial ensemble was to obtain a passage to the exterior through different climatic areas, and also to create a "membrane space". The building is therefore transformed into an activating organ for perceiving the changes of nature.

A special characteristic of this winter garden is its construction technology. The glass cube has a skin of insulating glass and a glued glass-only construction. The basic structure consists of two columns screwed to two beams at the front, constructed in triple-sheet laminated glass. The roof boards of isolated toughened safety glass with "fritted" ceramic patterns, casting 40% shadows, are glued to these beams. The minimal number of screwed connections is not part of the carrying structure, but has been caused by the necessity of faster drying-time of the silicon-adhesive (as the building was constructed in winter).

Another innovation is the frameless door of insulating glass, which is the first of its kind. This type of construction in which the glass is secured by adhesive is a technological innovation that had never been used in Austria, where the climate is quite harsh.

The use of a single material, glass, and the simple details of the structure give this high-tech assembly the appearance of a simple construction. Through this paradox the architects avoid the effect of "technical expressionism".

Photographs: Rupert Steiner / Archiv Eckelt-Glas

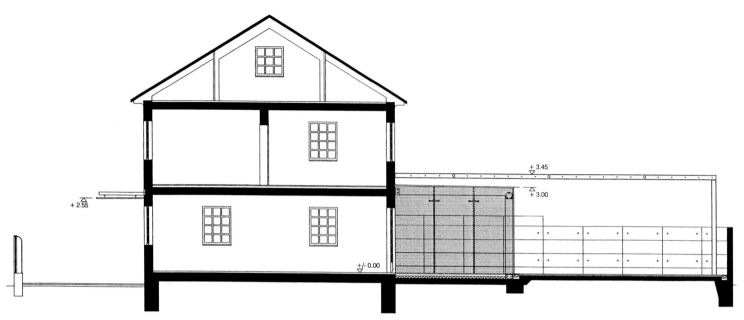

Cross section

The structure of the main entrance of the house becomes a curious, transparent room in which the aim is to emphasize the exterior beauty without fear of losing privacy. A smaller glazed structure is found at the rear entrance.

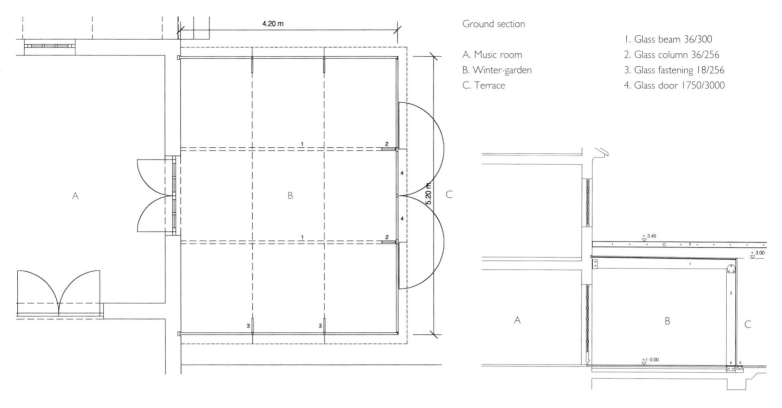

4.20 m

Ground section

A. Music room
B. Winter-garden
C. Terrace

1. Glass beam 36/300
2. Glass column 36/256
3. Glass fastening 18/256
4. Glass door 1750/3000

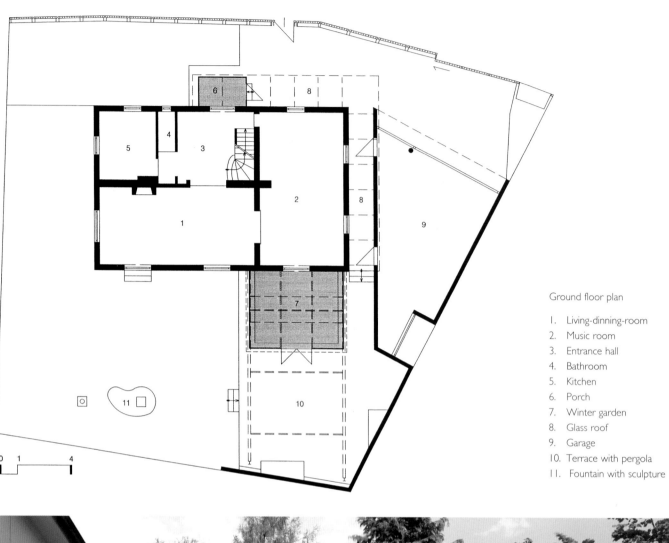

Ground floor plan

1. Living-dinning-room
2. Music room
3. Entrance hall
4. Bathroom
5. Kitchen
6. Porch
7. Winter garden
8. Glass roof
9. Garage
10. Terrace with pergola
11. Fountain with sculpture

0 1   4

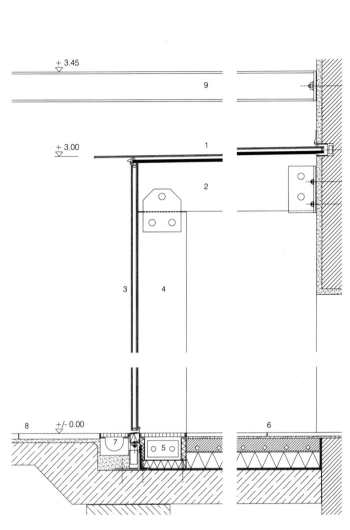

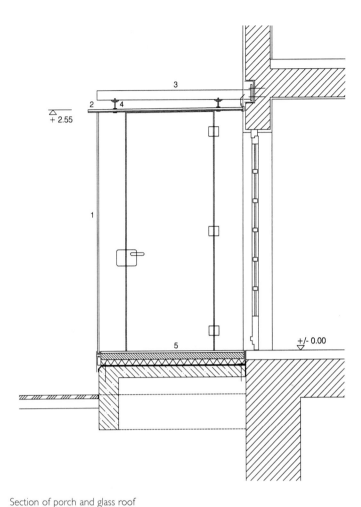

+ 3.45

+ 3.00

+/- 0.00

+ 2.55

+/- 0.00

Section of porch and glass roof
1. Toughened glass
2. Two 8 mm toughened laminted sheets
3. Galvanized steel-sword 125/20 mm
4. Point-fixing stainless steel

5. 16 mm mat of brush
70 mm plaster
50 mm roofmate insulation
Layer against humidity

Section wintergarden
1. Triple glazing, consisting
   of 8 mm layer of non-tinted glass,
   12 mm air cavity and two sheets of
   12 mm thick thoughened glass laminated together
   (with "fritted" ceramic point-pattern of 40%
   shadowing-capacity)
2. Glassbeam 36/300 mm
3. Double glazing 30 mm
4. Glass column 36/268 mm
5. Heating duct
6. Limestone paving 20 mm
   Bed of mortar 5 mm
   Plaster with underfloor heating 75 mm
   Insulation layer against humidity 80 mm
7. Prefabricated elements of drain
8. Limestone paving 40 mm
   Bed of mortar 30 mm
   Concrete 120 mm
9. Pergola HEA 160

Jestico + Whiles
Independiente

London, UK

After selling the record company Go!Disc to Polygram, Andy Macdonald looked for a home for his new company, Independiente. Although it was vacant and decaying, he recognized the potential of the building located at the south end of Turnham Green, near the Sanderson factory, designed by the architect Charles Voysey, and Bedford Park.

Since the original brief did not require facilities that provided a large amount of natural light, it was limited to a series of small windows in the north facade and three skylights. The encroachment of the surrounding urban grain has gradually worsened the already limited internal natural light, not least because of a massive brick building immediately to the south. However, the existing building retains many qualities. Behind its brick facade it hides a space articulated by a series of vaulted laminated arches, each of which is strapped by muscular, bolted iron ties. In danger of collapse through years of neglect, the timber bases were rotted from rising damp.

At first sight, it could be said that the client program of a series of acoustically separated spaces is at odds with the single volume of the existing building. Jestico + Whiles have organized these individual spaces around the perimeter of the building, creating a double height space. The skylights were reconstructed to provide good natural lighting that reaches the ground floor through the glazed blocks that form the access to the mezzanine. The building's proximity to its neighbors (in particular the large residential block immediately to the south) necessitated particular care in sound insulation. Reconstruction of the roof involved a significant increase in acoustic insulation, which is matched by the design of the ventilation systems.

Mixed-mode in design, the building can be heated and ventilated via a compact air-handling system, thereby isolating it from its surroundings. A small heat exchanger ensures that energy in use is minimal. In warmer conditions, the system can be used to supply air with the rooflights opened manually, allowing the natural stack effect to move air through the open area of the building. Throughout, the building's industrial history is reflected in the use of raw metal and low-cost industrial components.

Photographs: Paul Ratigan

The space was divided into two levels through the creation of a loft which is accessed by a staircase of glass blocks, a material that contributes lightness to a building dominated by solid wooden arches.

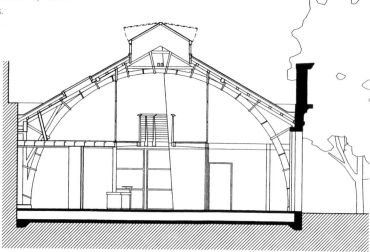

Cross-section.

Ground floor plan

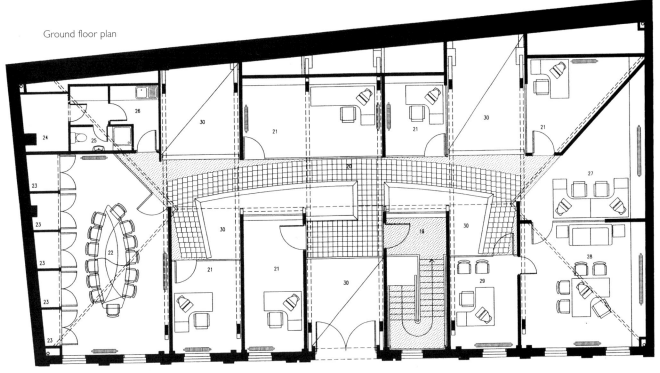

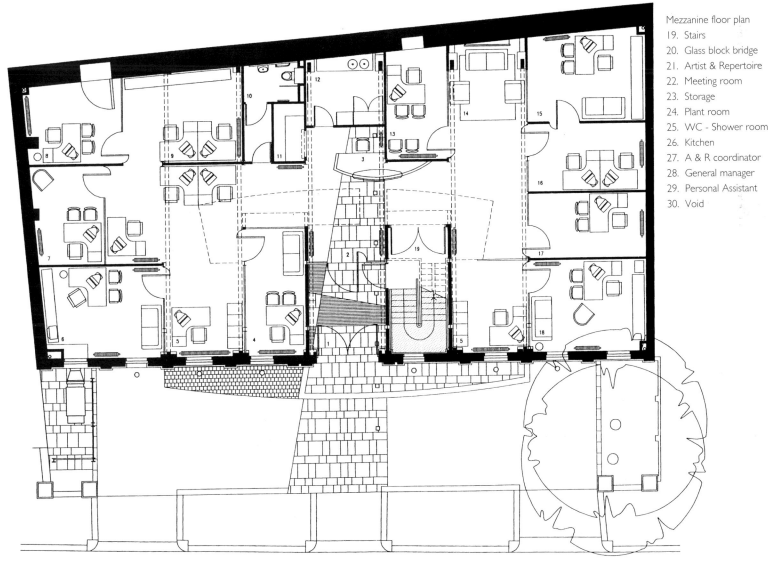

Mezzanine floor plan

316

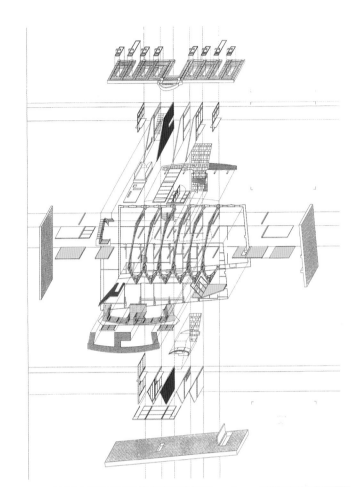

The space is defined by the interplay of geo-
metries and volumes. Oblique lines alternate
with vertical and horizontal lines, and all of
these with the curve of the arches that support the structure of the building.

Roberto Menghi
Castle in Lodigiano

Lodigiano, Italy

The aim of the project was to restore the north-western part of the Castle, modernizing the interior and making it fully habitable without spoiling the unique character it had taken on over the years.

New connections have been made between the ground and the first floors through the insertion of two new steel spiral staircases with steps in solid bay oak; the existing straight staircase in stone was resurfaced. The horizontal structures on the first floor, original beams and shelves in bay oak, were restored and reinforced with a special procedure based on resin and metal inserts. The existing terracotta tile floors were levelled off and integrated with new hand-made tiles of the same size and in the same clay (which is still found in the area) as the old ones. To solve the problem of rising damp, which has been penetrating the wall for centuries, an insultation system using "active electro-osmosis" has been adopted with excellent results.

The project also creates an intermediate level between the ground and the first floor, in part adapted as a study-library facing over the main hall, in part used as a service area: cloakroom, laundry, etc. The six meter ceiling height typical of the age has been maintained for the entrance hall, part of the main hall and the whole kitchen. The intermediate level has been made, like the stairs, using steel structures, but with flooring in Swedish pine.

The two stairways and this last space were designed and constructed with materials that contrast with the original context, in order to highlight their super-structural and "removable" nature. Bedrooms, bathrooms and closets have been fitted into the space situated above the pointed arches. The outer wall of the bedrooms overlooking the courtyard has been moved back from the rest of the façade in order to make room for a long (approximately two meters in width) balcony, above the arches.

The Castle looks like a fortress with a moat, having surrounding walls and a shortage of apertures, with a consequent shortage of light and air in the rooms. This problem was solved by reopening some of the original apertures that had been walled over and making some new horizontal slits under the western eaves.

The roof has been restored and insulated, maintaining the existing cover in bent hand-made tiles.

Photographs: Melina Mulas

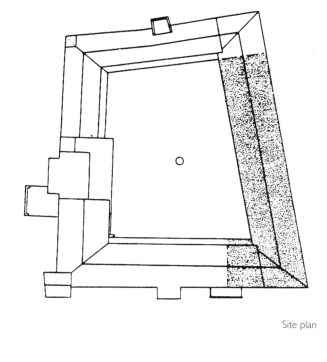

Site plan

View of the large courtyard of the castle, with large porticoes providing spaces protected from direct sunlight.

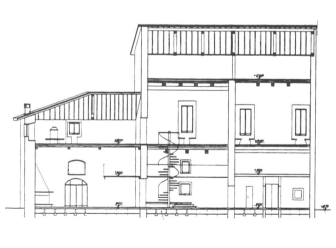

Section A-A

Ground floor plan   Intermediate level   First floor plan

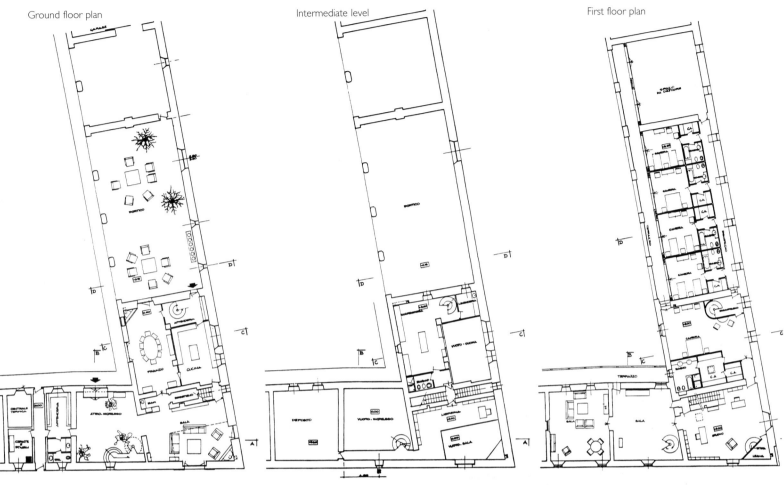

Part of the work consisted of inserting a new level between the ground floor (6 meters high) and the upper level, and establishing a new vertical connection by means of a new spiral staircase with solid oak steps.

The original loadbearing structure made of wooden trusses has been treated with resins for conservation and reinforced with metal elements.

Stig L. Andersson
Andersson's Apartment and Architectural Studio

Copenhagen, Denmark

The flat, located in central Copenhagen, is inside a block dating from the early 20th century. It contains the residence and studio of the Danish architect Stig Andersson.

The restoration focused on intensifying the experience of the textural effect of the surfaces. "Reason is not interesting. It is all about 'contemplation of the surfaces', a meditation. (...) One learns from looking into walls", says Andersson, a great admirer of the Japanese tradition.

"My wish was to create a home in this spirit. Over the years many people have lived in this flat. More than seven wallpaper designs were the signs of this long story. By removing all the superfluous designs I have achieved an impression rich in experiences.

"So I decided to leave the walls entirely free of wallpaper, only asking the craftsmen to fill out some of the irregularities and gaps. This is the rawest impression one can create. Only where necessary I took a bit more care, for example on the walls where all the pictures are placed. The refined lines appear more clearly against the rustic impression of the background. By placing more refined objects on the rough walls the items look more elegant and delicate".

The colors in the flat are very vague. "In our country one will find many shades of gray, which change with the light all the time. The Danish landscape has many fine shades. There are no intense impressions to see as in the south."

Very strong colors seem strange in Denmark; they are better in stronger light. "For me," says Andersson, "dusty colors are important. They give me a feeling of delicate brittleness. It is the matte impression I was investigating. The diffuse light makes things look as if they are floating, making them lighter". The project combines sand colors ("very important for my well-being", declares Stig Andersson) with elements in wood. "Wood has the natural color of the material. The surface can be changed by coarsening or lacquering the wood".

All these thoughts came to the architect's mind through observation of the Japanese tradition. "In Japan, they carry sophistication to the extreme. One can see this by looking at old teahouses, the lacquers, ceramics, woodworks and textile designs". A long corridor that links all the rooms is typical in this kind of flat in Copenhagen.

The rather dark space of the corridor of this house gets light from a row of low hanging bulbs, which give a warm and intense light. At the same time, the asymmetrical placing of the lamps makes the corridor seem wider.

Photographs: Ostbanen

The living/dining room is dominated by the considerable presence of a large, light-hued wooden table with steel legs and a matte finish.

The architect has opted for bathing the spaces in a diffuse, rather undefined light. By avoiding bright colors and high contrasts, the objects seem lighter and appear to float in their surroundings.

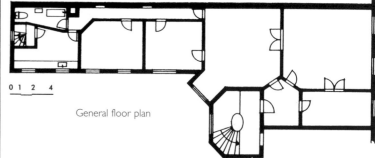

General floor plan

0 1 2 4

The project experiments with the effects produced by different textures and surface finishes, while maintaining the irregular and rough look of the walls.

Ottorino Berselli & Cecilia Cassina

Residence in Puegnago del Garda

Monte Acuto, Italy

In the Italian hills of Valtenesi, overlooking Lake Gada stands the small town of Monte Acuto. It is presided over by a tower with an old dovecote, a construction typical of the plain of Padania but uncommon in this region.

The restoration project covered one corner of the old town, which dates backs to the end of the 6th century and is dominated by the massive quadrangular tower. The whole complex was is an advanced state of disrepair, particularly the tower, and a thorough restoration was therefore carried out on several levels.

The project was based on the idea of retrieving previously existing elements, with particular attention to interpreting everything that, through the successive layers that had been superimposed over the years, bore clear indications of its past. The project therefore proposed a reassessment of the whole town, despite the fact that it directly affected only a fragment of it, "in an attempt to regain lost urban emotions", in the words of the authors in their report on the project.

The theme running through the new work is light, which assumes the main role in the scheme as a result of the generous yet subtle openings. The serene atmosphere of the interior of the old tower (now converted into a dwelling) is bathed in light entering through clean slits running the length of the ceiling on the first two floors, where the main rooms are situated.

The material used – palette-applied intonaco on the walls, and colored cement and natural oak on the floors – endow all the rooms with a homogenous feel. The distribution of the rooms was problematic due to the unusual structure of the house, particularly the top story, formerly used as a dovecote. Finally the solutions adopted were to use the ground floor for the bedrooms. The remaining floors (the second and third floor and the former dovecote) contain the guest rooms, various study areas and a library with panoramic views of the lake.

Photographs: Alberto Piovano

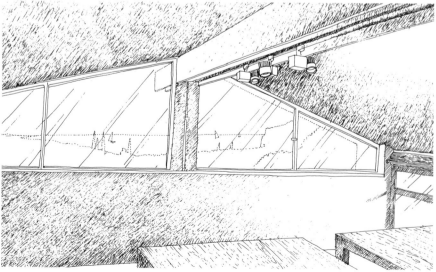

North elevation

Section AA

Section BB

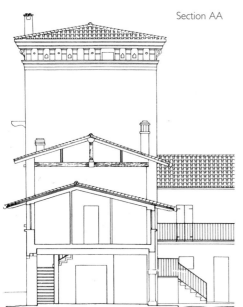

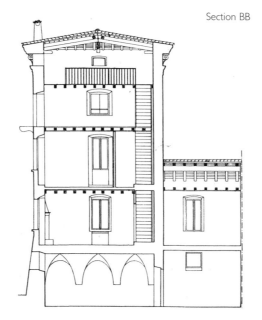

The project involved the restoration and renovation of an old building featuring a tower crowned with a dovecote, now converted into a dwelling. Before the reparation the complex was in an advanced state of disrepair.

Ground floor plan

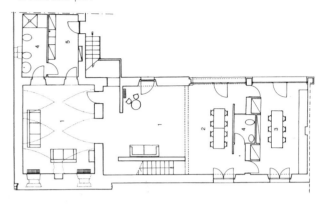

First floor plan

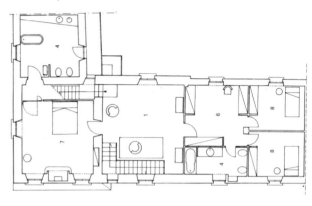

Second floor plan

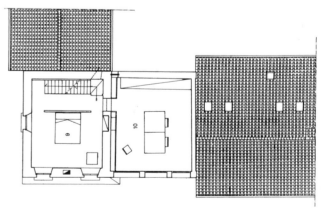

Third floor plan and dovecote

 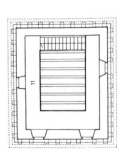

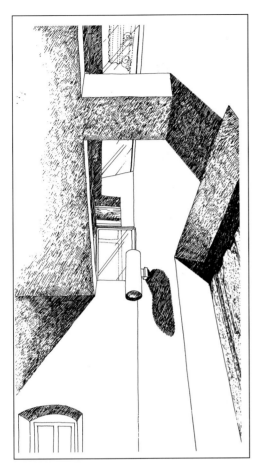

Luis Vicente Flores
X' Teresa,
Alternative Art Center

Mexico City, Mexico

The fundamental objective of the Santa Teresa la Antigua project was to transform the existing building into a space for the presentation and staging of non-conventional art that requires architectural conditions that are autonomous and independent from the original structure.

Due to their dimensions, the existing spaces can be adapted to the new requirements with relative ease. However, due to the modifications that the building has undergone throughout its history, it lacks a coherent spatial structure that relates the different rooms, and above all it lacks a sequence that articulates the whole from the access to the different rooms and the open areas. It was necessary to create a flexible space with respect to the presentation of the works and the circulation of the public. Because of this it was proposed, firstly, to eliminate the elements separate from the original structure and the restructuring of the space from a new access point that allows the circulation to be organized. An extension will also be designed in the rear courtyard, with the aim of concentrating vertical circulation and housing services that will disappear when the existing rooms are gutted; it will also increase the amount of office space.

To reorganize the spatial sequence, it was proposed to create a new access by means of a ramp through one of the windows of the main façade. The sequence of circulation is thus substantially improved and an optimum utilization of the main room (the transept) is enabled through the removal of the existing accesses.

The extension is a light steel and glass element annexed to the existing building without a structural relationship. The location of the element added within the rear courtyard was used to convert it into a three-dimensional support for installations and performances in the open air. It is conceived as an abstract element solved through transparent lines and planes. Due to its geometry and lightness it will contrast with the volume of the existing building. Structurally, it was designed as a detachable body that can be understood as provisional.

Photographs: Pep Avila

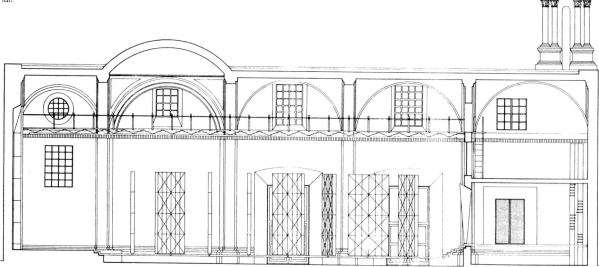

Longitudinal sections

Ground floor plan

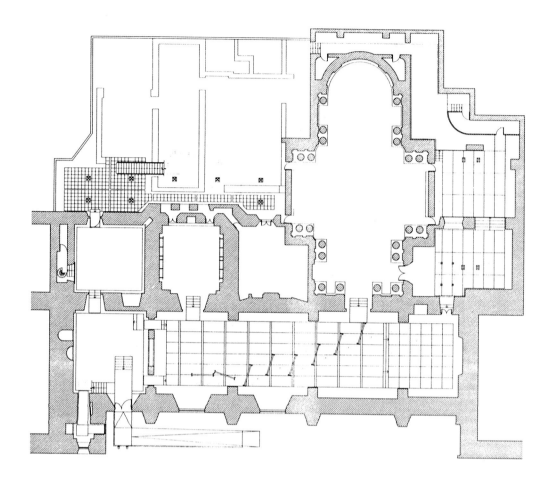

Upper floor plan

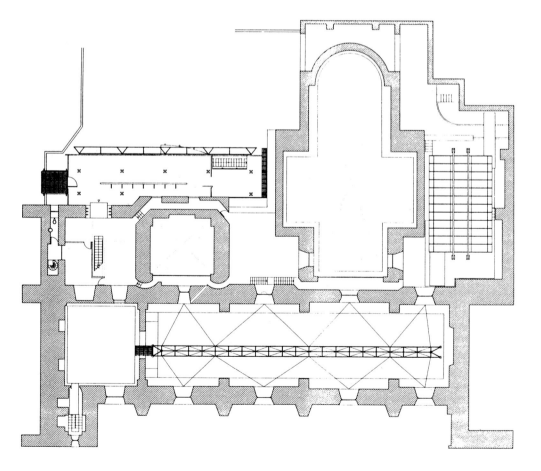

The collapse of the east side of the X' Teresa center caused considerable structural deformations.

In order to take full advantage of the possibilities of the building and to guarantee maximum flexibility for displaying the works of art, all the interior partitions of the original structure were removed.

Cross sections

The extension to the building was added at the rear, looking onto an inner courtyard. It was thus possible to concentrate the vertical circulation and increase the amount of office space.

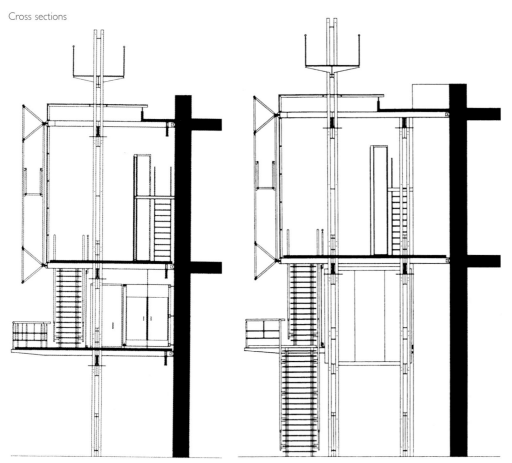

Courtyard elevation

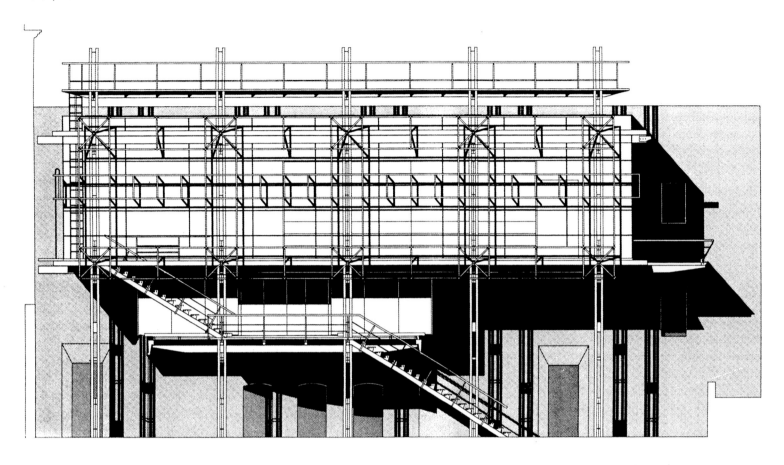

Alois Peitz
St. Maximin Sports and Cultural Center

Trier, Germany

The three main requirements of the brief to restore St. Maximin church were its conversion for accommodation of sporting and cultural activities, access to the archaeological remains beneath the church and retention of the newly restored appearance of the building both internally and externally.

The main space is reached via an extension of the former vestry, in which a foyer and the ancillary spaces are located. The wood sprung floor necessary for sporting uses lends the tall, hall-like space a warm note. Wall bars and basketball nets have been fixed to the walls of the side aisles. To protect children against injury (and the structure against damage), temporary matting is fixed to the columns during sport sessions. From an overhead gantry nets can be lowered to divide the nave into separate sections without interrupting the visual continuity of the space.

The lighting concept provides bright, cool light for sporting activities and warmer light for concerts and other cultural events. Spotlights directed to mirrored surfaces produce lighting without glare. The orchestra platform in the former chancel can be extended by means of a hydraulic lifting system. Special plaster was applied between the vaulting ribs to improve the spatial acoustics.

The newly inserted elements—windows, doors, sports equipment, the gantry, the spiral staircase in the tower, and the lighting fittings—are mainly of steel. They accentuate the rigorous architectural language of the former church space and introduce comparable modern forms without seeking historicist clichés.

Photographs: Trapp, Oberdorf & Peitz

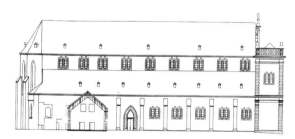

North elevation

Cross-section

One of the requirements of the
rehabilitation programme was to
provide an independent access and
to organise a route through the
archaeological remains in the lower
part of the church.

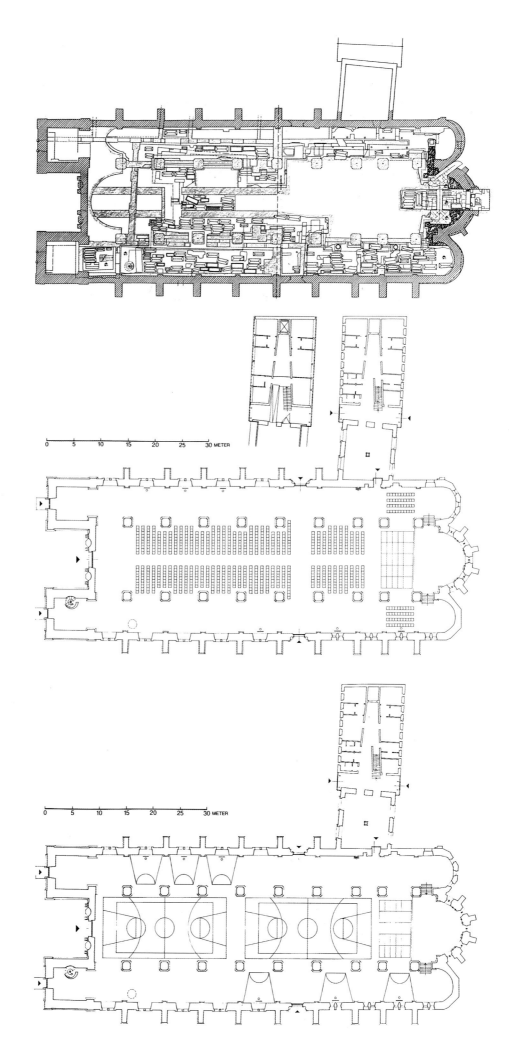

Basement floor plan

Plans of the church after the intervention

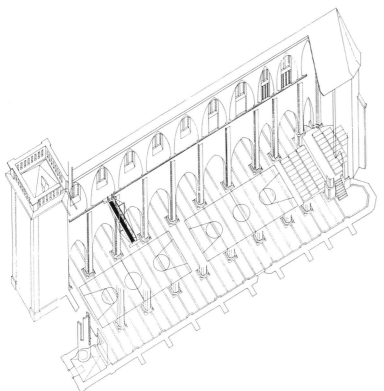

As can be seen in the photograph on the right, a system of nets descends from the upper part of the church, allowing it to be divided into different sectors without interrupting the spatial continuity.

The top photograph shows how a system of spotlights directed to mirrored surfaces provides an even quality of lighting in the whole interior.

One of the elements added to the church is the spiral staircase that leads the visitor to the top of the tower.

Claesson Koivisto Rune
Architectural Studio

Stockholm, Sweden

For their own office, Claesson Koivisto Rune sought for a shop-space in a side street in central Stockholm. They thought that being in contact with the street and the ongoing life of the city was essential for their inspiration and for their availability to clients The architects divided the long and narrow space according to hierarchy. The front room, which opens to the street trhough a display window and a glass door, became the meeting room. Between this room and the next, two large double-etched glass sheet walls were placed. They let outside natural light in, and one can pass between them on either side, but they obstruct the vision into the next room. From outside one can sense the continuing space, but not what it contains and not how much there is. There is also one floor glass letting down to the room beneath. Behind the two glass sheets fluorescent lighting was installed.

The tubes are UV-coloured and turned on at night as a light effect. Next room became the hub of the office. Computers, faxes, telephones... Inside the "hub" are the lavatory, the kitchen with a small dining area and the stairs down to the lower floor. Downstairs, the library, drafting boards and storage rooms. Last, under the front room, the model workshop.

The background to the design had to be neutral, and thus white was the colour chosen for all the office and the furniture.

In the office there are no doors (except for the storage rooms and lavatory), but the division into rooms/functions is clear. The disposition of the different spaces allows the artictects to choose how long into the design process they let their clients, maybe just the presentation room, maybe all the way into the model production in the workshop. The room hierarchy is also one of order. Design is a messy business, especially model making.

Photographs: Patrick Engquist

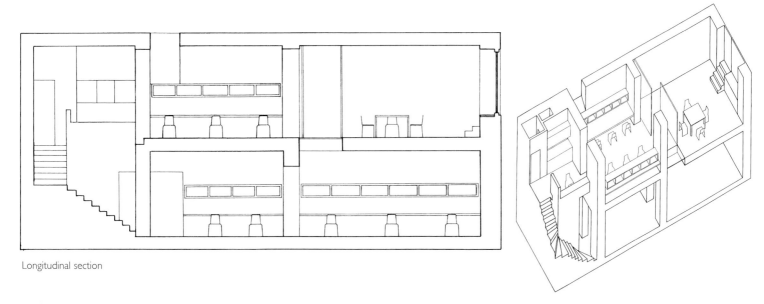

Longitudinal section

At night, the fluorescent tubes behind the translucent glass panels that separate the reception from the interior of the study reinforce the image of the outer room as a shop window. Behind the panels is the "nucleus" of the study, the production area.

Ground floor plan.

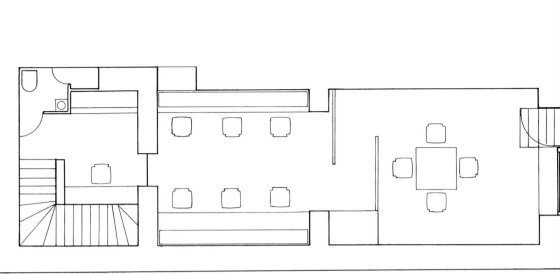

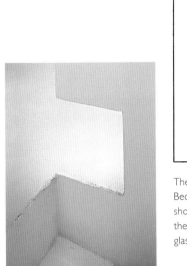

The studio is accessed directly from the street. Because the premises were formerly used as a shop, the reception, like a shop window, connects the interior and the exterior. The window and the glass door provide natural light to a semi-basement.

The space is defined by geometric forms and the absolute predominance of white. The audacious use of lighting as a decorative element reinforces the design concept of the scheme.

Massimiliano Fuksas Architetto
Studio in Rome

Roma, Italy

Fuksas Associati's studio occupies three floors of a 16th-century building located in the Piazza del Monte della Pietà, in the heart of Rome. The exterior of the building is not the reddish colour characteristic of the houses in the city, but has been painted pale blue. Wooden doors are used for the entrance, on the first floor. The office gives a rather chaotic welcome to the visitor: original elements cohabit with later interventions that are given less emphasis. A glass wall separates the entrance hall from a meeting room and leads to the secretarial area without modifying the original dimensions of the room. The most significant characteristic in this office is its walls: although they are decorated in very different ways, all seem to exhibit abstract images on their surface. The former occupants of the house have left their marks over the centuries by means of layers of colours on the walls; these were sanded and polished until the surface was smooth, and finally treated with wax. Thus, no wall is the same as another, each one has its story to tell. An interesting case is a room that the former owner used as a toilet: after removing different layers of paint, the frescos of an old private chapel were discovered. This way of revealing history demonstrates the close relationship of the architect with Rome, his city, the city of strata as it is now defined.

The constant presence of the past does that new proposals are absent: the glass lift that breaks through the wooden beams of the ceiling in another room is a good example of the coexistence of new and old. This lift leads to the second floor of the studio, where most of the forty employees work, divided into groups according to projects. Here, walls subdivide the study into smaller spaces with glass doors on steel hinges.

Each room seems to be the antechamber of the previous one, there are no hierarchies or indications to guide the visitor. Even the furniture seems to have found its place in a fairly random way. All the work positions have a computer, but their location is not fixed. The third floor houses the most silent environment of all. It is accessed by a smaller elevator that also leads to the roof. As on the first floor, here the windows are large, and the light that penetrates into the interior through the windows transforms the walls into striking reliefs.

The exterior of the building is not the reddish colour characteristic of the houses of the city, but has been painted pale blue. Wooden doors mark the entrance, on the first floor.

The inner courtyard and the worn staircase of this 16th-century palazzo suggest the entrance to a dwelling when in fact they lead to an architecture studio.

On the upper floor, the wooden structure that supports the roof has been left exposed and painted the same colour as the walls. The combination of the original structure of the building with the furniture of the office, some of which was designed by the architect, creates a pure and dynamic environment.

Photographs: Giovanna Piemonti

The first things that the visitors notice in the secretatiat are a painting of the architect and posters from exhibitions of his work, covering the historical walls like a collage. A glazed elevator leads from a conference room on the first floor to a room above it equipped with a plotter, a server and a photocopier.

Right: A glass lintel divides the entry area from a conference room. The panes are held in place with filigraine sash angles made of steel.